From

AGAR
to
ZENRY

A book of plant uses, names and folklore

RON FREETHY

With illustrations by
Carole Pugh

The Crowood Press

BY THE SAME AUTHOR

The Making of the British Countryside (David & Charles, 1981)
How Birds Work (Blandford, 1982)
The Naturalist's Guide to the British Coastline (David & Charles, 1983)
Man and Beast (Blandford, 1983)
British Birds in their Habitats (Crowood, 1985)

First published in 1985 by
THE CROWOOD PRESS
Ramsbury, Marlborough,
Wiltshire SN8 2HE.

British Library Cataloguing in Publication Data
Freethy, Ron
  From agar to zenry: a book of plant uses,
  names and folklore.
  1. Botany, Economic
  631   SB107

  ISBN 0-946284-51-2

Design by Vic Giolitto

Phototypeset by Quadraset Limited, Midsomer Norton, Bath

Printed in Great Britain by Robert Hartnoll Limited.

Contents

Acknowledgements

My thanks are due to Brian Lee, who read the proofs with great attention to detail, and to my wife, Marlene, who typed the manuscript and whose deep botanical knowledge was of constant help to me in writing this book.

To the Reader

It is always tempting to view books about herbal remedies as a substitute for the medical profession. Whilst many plants are harmless as well as useful, an accurate dosage is often crucial and the correct identification of a species or of the parts that are non-toxic is frequently not as simple as it appears. If, therefore, you are tempted to try herbal cures yourself, you should only do so with the guidance of a skilled herbalist or homeopathic physician – and you should never attempt to treat illness without consulting your own doctor.

Even common plants can be deadly, so if you wish to be adventurous in what you eat, before you experiment consult a reliable reference book such as Pamela North's *Poisonous Plants and Fungi* or the *Hamlyn Guide to Edible and Medicinal Plants of Britain and Northern Europe*, details of which are to be found in the bibliography.

It should also be emphasised that many of the plants described in this book are uncommon enough to be protected by the Wildlife and Countryside Act – and care should be taken not to damage even the commonest of plants or trample the area around them.

From Agar to Zenry is not a formal history or a comprehensive reference book. Indeed, the uses of plants and trees and the origin of their names is such a vast subject and so much has been handed down in the way of traditional wisdom and folklore that even the most dedicated naturalist or scholar can never hope to master more than a fraction of all there is to learn. What I have attempted to do in these pages is to set down some of the 'green cunning' imparted to me by my great-grandmother – some of her crafts, cures, recipes and wisdom – adding to it some of the knowledge that I have gleaned for myself as a naturalist and scientist and pointing out wherever relevant how the plants that were useful to our ancestors can still be useful today.

Nowadays the woodland crafts that not so long ago still existed are in danger of being forgotten. It would be a pity if knowledge of these crafts was lost altogether. I have therefore included brief descriptions of trades and crafts such as bodging, clog making, thatching and charcoal burning, many of which have already vanished, in the hope that readers may be encouraged to learn more about them. Finally, for the naturalist I have included items of botanical information that are of interest or which may be of use in identifying the species described.

Ron Freethy

1 Green Cunning

In the old house on the coast where I grew up we always seemed to have a kitchen full of bounty from the sea. This was due entirely to my great-grandmother. She was eighty-nine when she came to stay with us in Cumbria, brown as a berry, wrinkled as a nut and fit as a butcher's dog. Her West Country accent amused the villagers and because of her wanderings on the beach and in the woodlands, plus her diminutive size, the local kids called her, not without affection, the little witch. I was almost six when the war came, and all through the long winter of 1940 we sympathised with each other, as the supply of sweets in the shops dwindled, for she loved children – and loved sweets as much as we did.

'Thee wait till summer, dearies,' she grinned, her brown eyes twinkling with mischief, 'an' I'll show 'ee summat.'

As the driftwood gathered from the storm-strewn beach crackled in the old black-leaded grate and the gas light flickered and hissed in its mantle, what tales she told! She said we would go short of nothing – so long as the sun shone and the grass grew. The snow stayed late in 1940, but by the end of March the pair of us were out searching the south-facing slopes of hedgerows and woodlands. Throughout the long, hot summer we wandered sea and sand, dune and hill, field and farm, hill and dale. On her ninetieth birthday she was up to her knees in a ditch picking watercress. Neighbours who did not believe in witches thought her harmlessly mad, for who else would festoon a house with wild plants hanging

on strings like onions, fill old shoe boxes with fruits and seeds, gather seaweed from the beach (rumour had it that she ate it) and gently strip bark from a tree so as not to kill it, so she could do the same next year? But gradually things began to change. First, she gave one of our neighbours feverfew for her 'vapours'. Then, when potatoes were scarce, she persuaded us to eat mashed silverweed and Auntie Doris on a surprise visit tucked in and failed to detect the difference.

Blackthorn leaves added to the pot made tea go further and, some said, improved the taste. Local babies short of orange juice thrived on her home-made rose hip syrup – and I liked it too, especially diluted like a cordial. She brought dogs, cats and a farmer's pig back from the brink of death with knitbone juice and would never take a penny for her trouble. 'It's God's gift to us all, is green cunning'

Blackthorn

was her only reply. Then there were the sweets. Coltsfoot rock, snow eringoes, Spanish water and wild strawberries were just a few of the unfamiliar delights to reach our palates. The adults were equally pleased with the little witch's yeast culture, which bubbled and frothed when added to potato peel, parsnips, nettles, every wild fruit imaginable, oak leaves, dandelions and even grass. When corks popped merrily in the evening and wines and cordials made from leaves or berries helped to digest the meal of cockles, mussels, winkles and whelks, the old lady who provided the feast put it all down to 'God's gift to them with animal cunning'.

The little witch died in her one hundred and fourth year with a quart jug of bilberry wine by her bedside and took with her a store of green cunning passed down by word of mouth, mostly from mother to daughter. Without her interest in plants my own love of flowers and folklore would never have blossomed. This book contains much of her green cunning. It is my own small tribute to her. Most of it was written in the open air, where her words seem most vivid and

Wild strawberry

her world is all around me. From her I know about agar, made from seaweed, and about zenry, the old West Country word for charlock.

2 Plants in History

Before iron spears and the bow and arrow were invented, man lived mainly on plants. Fresh vegetables supplemented by juicy insects and worms would have been pleasant enough during the warmer periods and in times of plenty, but in times of scarcity and in the colder periods, when survival was a problem, great reliance must have been placed upon stores of fruits and seeds and on a detailed mental map of where to dig for bulbs, corms, rhizomes, tubers and tap roots. A knowledge of plants was therefore essential for survival and it was the great botanists of prehistoric times who survived to pass on their knowledge to their offspring and to future generations. They learned by trial and error which plants were non-poisonous and ate, and possibly even enjoyed, plants which we would never think of eating today. This fact was appreciated by the Roman natural historian Pliny, who wrote in his *Naturalis Historia* that 'it is impossible sufficiently to admire the pains and care of ancients who explored everything and left nothing untried'. Indeed, as a result of his readiness to try everything, early man advanced quickly in his scientific investigations, perilous though they were, into what was edible and what was not. Although the uses of plants for food no doubt loomed large in the mind of early man, it would be a mistake to underestimate the part played by plants in other aspects of human evolution. Timber was widely used in the construction of dwellings and other buildings, early writing was inscribed on decolourised leaves and

papyrus, and thousands of useful artefacts, including boats, carts and wagons, bridges and farm implements, all demanded a detailed knowledge of trees. The earliest use of plants, however, was undoubtedly as food. Gradually there also developed an appreciation of plants which, although not nutritious in themselves, gave added flavour to food and made it more palatable, while other plants were used as preservatives and still others as medicines.

In the early days the human species was nomadic and it was only around 10,000 BC, in western Asia, that the idea of cultivating crops developed, ending the need for continual and often fruitless searching. By 6,000 BC Neolithic societies were quite well organised in Europe and Asia and by 4,500 BC the ox-drawn plough was a feature of a more developed society, with an increasingly scientific approach to improving wild plants. Across the Atlantic the cultivation of pumpkins and peppers may well have been in full swing by 6,000 BC and the natives of Central America were successfully growing maize by 2,500 BC.

A great deal of Palaeolithic art portrays animals hunted for food and these cave drawings often show intricate and accurate details. But plants appear seldom, and even then only as background, without fine enough detail to determine the species depicted. This is probably a case of familiarity breeding contempt, but there is an Egyptian mural dating from 3,200 BC or earlier which depicts cereals (probably wheat and barley) and date palms. From this time onward more and more emphasis was placed upon plants and their cultivation in field and garden. Indeed, Egyptian herbalists were famous and in Homer's account of Helen of Troy he mentions that the fair lady obtained a supply of drugs from an Egyptian woman. Dictionaries of useful plants were prepared by the Egyptians and by subsequent Asian and European civilisations, but the boundary between botany and religion was very fine indeed. There was a parallel development in China, which has retained rather more of its herbal traditions, because herbalists in China have suffered very little from religious persecution.

When Christianity set about ridding the world within its influence of the old religions the priests retained some of their knowledge relating to herbs but frowned upon the ancient customs when they were practised by others, though the knowledge of the more practical uses of plants in such industries as building and dyeing and in the manufacture of artefacts remained untouched. In time, however, ordinary country folk began to be persecuted or put to death for continuing to dose themselves and their families with simple, and often effective, herbal remedies. Witchcraft and sorcery are harsh names to apply to such simple practices. During the period of the crusades the Knights Hospitallers, who belonged to many nations, treated the wounded and sick using their own native folk remedies and also those they had learned from the Saracens. They therefore acted as a collecting point for the knowledge of several civilisations. The Arabs had a wonderfully erudite knowledge of herbal medicine, but Christians regarded them as heathens and therefore blasphemers and fear of the so-called black arts was fostered by the Inquisition, who ruthlessly hounded the otherwise devout Hospitallers, accusing them of being followers of Satan. Even today herbalists still have not managed to rid themselves of this unwarranted stigma, although their cause is not helped by the occasional charlatan who continues to prey upon the susceptible and the superstitious.

In the Middle Ages most of the population lived in the country rather than huddled in towns. As a result, even the youngest children knew which wood was best for firewood, furniture or axe handles, and almost every cottage had its own herb garden to stock the cooking pot or cure minor ailments. A diminishing number of country folk still possess these skills and have sufficient knowledge to prevent frequent visits to doctor or chemist.

Many people imagine that the rarer the plant, the greater its virtue. However, for our ancestors many common weeds were invaluable items in the medicine chest. The dandelion, for example, is usually regarded as a nuisance, but it was frequently and successfully employed as a health-giving remedy for complaints of the liver, kidney and blood and I have often talked to country folk who insist that dandelion eases rheumatic pains. The nettle is hated by those who have felt its sting, but the young tops, gathered in spring and eaten as a boiled vegetable, were regarded as an excellent blood purifier. Raspberry leaves, especially those of the wild plant, were, when infused, highly regarded for treating 'dirty tongues' in infants, and the same concoction when taken by expectant mothers was reckoned certain to prevent miscarriage. Eyebright, that small but familiar plant of grassy and upland pastures described in Chapter 8, was used in the treatment of sore eyes. The scientific name for the species is *Euphrasia*

Wild raspberry

officinalis. The latter word indicates its value to the old apothecaries, *officina* being the Latin for an apothecary's workshop. Eyebright flowers have a central area which seems to gleam like a bright, healthy eye – a perfect example of what has become known as 'the doctrine of signatures'.

From classical times (Aesculapius, the Greek god of medicine, was said to be skilled in the medicinal power of plants), the study of botany was important to the art of healing. The language of medicine, however, was far from precise even as late as the seventeenth century, as is illustrated by Nicholas Culpepper, the leading herbalist of his period, who described himself as a 'Gent, student in Physick and Astrology'. Here is what he wrote about bugle, which grows commonly in woodlands:

'This Herbe belongeth to Dame Venus. If the vertues of it make you fall in love with it (as they will if you be wise) keep a syrup of it to take inwardly and an ointment of it to use outwardly alwaies by you. The Decoction of the leaves and flowers made in wine, and taken dissolves the congealed blood in those that are bruised inwardly by a fall or otherwise. It is very effectual for any inward wounds, thrusts or stabs in the body or bowels.'

Such a plant would obviously have been a vital ingredient in the herb collection of Dioscorides, the famous soldier-doctor in the days of Nero. He and the many who followed him held that every herb was under the dominion of either the sun or one of the planets, which could be discerned by the signatures or outward appearance of the plants themselves. The plants of Saturn were those with hard dry leaves, rather dull flowers and not too pleasant odours; those under the control of Venus, on the other hand, had foliage of a rich, succulent green, the flowers being bright and sweet-smelling. It was

Bugle

Greater celandine

their Vertues, as the learned Grollius and others will observe; as the Nutmeg being cut resembleth the brain; the Papaver erraticum, or Red Poppy resembleth at bottom the settling of the Bloode, in the Plurisie, and how excellent is that Flowre in diseases of the Plurisie and Surfeits has sufficiently been experienced.'

According to Culpepper, the deep yellow colour of the sap of the greater celandine was a sign that it is a specific for jaundice, as were other bright yellow plants: 'Behold here another verification of the learning of the ancients . . . that the vertue of an Herb may be known by its signature, as plainly appears in this: you shall perceive the perfect Image of the Disease in some part of the plant.'

Even a casual glance through an Elizabethan or Stuart herbal reveals that these were indeed violent days. Plant

also believed that the various parts and organs of man and animal were under the control of particular planets. The doctrine held sway for many centuries and an echo of it still exists when we refer to people as jovial, mercurial or saturnine.

According to the doctrine of signatures, a plant bearing a particular signature could be used to treat an organ of the body bearing the same signature, as explained by Turner, Dean of Wells at the time of Elizabeth I, who produced the first British herbal:

'God has imprinted upon the Plants, Herbs, and Flowres, as if it were in Hieroglyphicks, the very Signatures of

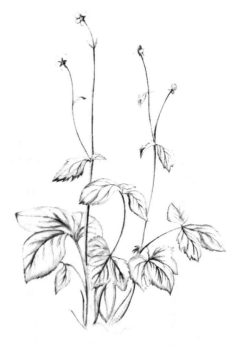

Herb bennet

Mandrake

extracts which would ease the pain of cuts and bruises, often inflicted by sword or bludgeon, or prevent wounds becoming septic were in great demand. These were classed as 'simples', made up of extract from only one plant and distinguished from 'compounds', which consisted of several extracts blended together.

John Gerard, a barber-surgeon and perhaps the best known of the seventeenth-century herbalists, is full of plants guaranteed to ease the pains resulting from a fracas. His work is often spiced with humour: 'The Roote of Solomon's Seale taketh away in one night or two at the most any Bruise, black or blewe spots, gotten by falls or Womens wilfulnesse in stumbling upon their hasty Husbandes Fistes.'

Solomon's Seal, a handsome but not common plant found in limestone woodlands, was Culpepper's choice to combat mild disfigurements: 'The Decoction in Wine . . . or the distilled water of the whole Plante, used to the Face and other parts of the skin, cleanseth it from Morphew, Freckles, Spotts or Marks whatsoever leaving the face fresh fair and lovely; for which purpose it is much used by Italian ladies.'

Another woodland herb used in early medicine was herb bennet (*Geum urbanum*) or *herba benedicta*, which means 'the blessed herb'. Also known as wood avens or common avens, it is common in woods and hedgerows during June and July. Up to the eighteenth century it was employed as a nostrum against pain and as an antidote to any form of poison. An extract from the roots was used as a tonic and as an astringent.

Turner, Gerard and Culpepper were leading lights in the honourable profession of apothecary, but as in any lucrative calling there was no shortage of charlatans as well as those who exploited the superstitions of simple folk. Quackery and witchcraft took advantage of such popular superstitions and both made use of mandrake roots, which bear an uncanny resemblance to the male reproductive organs. Mandrake was thought to prevent sterility in both man and animals and was believed to be a powerful aphrodisiac. No wonder country folk called it Satan's apple and were terrified to dig up the roots. Instead, they tied a starving dog to the plant and placed a dish of food a short distance away. When the dog wrenched the plant from the ground, the mandrake was said to scream 'like the Devil', but once pulled from hell it did no harm. It was thought to grow particularly well under gallows and was also known as the gallows plant, but in England it has never been particularly common and bryony was often used as a substitute.

In those days, and even in the not too distant past, woodland provided timber for building and furniture, bark for tanning leather, and materials for a host of other essential articles. The woodland craftsman knew how best a particular tree would grow in response to light and wind and how to select the right timber for the right job. He worked at peace with himself and the world, but knew that if he made the wrong choice he would have to do the whole job again, as well as having wasted valuable wood. Carpenter, herbalist and apothecary all had this 'wort cunning', which allowed them to select each plant for a particular purpose, no matter where it grew – in hedgerow, forest, meadow, pond or stream, or on the seashore, moors or hillside.

3 Plants of the Sea

I eased the *Little Witch*, the boat I bought with money left to me in my great-grandmother's will, into a sheltered bay and dropped anchor as the tide began to flow in, allowing the buoyant seaweed to float. Most folk's idea of seaweed is of a slippery treacherous mass on a rock or a blackened heap of stinking vegetation tossed aside by a rough sea. Slimy, smelly things they are in these conditions; but seen from a boat, their delicate transparency highlighted by splashes of penetrating sunlight, they look like the world's most beautiful plants. Yet they are useful too, and some of them are essential in commerce. There are a great many species, classified according to colour, and all are seen at their best when you join them under water. The large and sinuous seaweeds which wave about in the currents can easily entangle and drown the unwary swimmer, so I always carry a sheath-knife strapped to my belt, just in case I need to cut myself free, but it is also useful for cutting specimens for use or identification. The waters deepening around the *Little Witch* as the tide flowed in were quite chilly as I slipped over the side and swam down to an underwater rock garden. The dominant brown seaweeds had floated clear and I could see their huge holdfasts clinging to the rocks, as if held by a powerful adhesive. It has been proved recently that the holdfast produces an acid which, despite being immersed in seawater, is powerful enough gradually to dissolve the rock, allowing the holdfast to grow into the crevices thus created. Other experiments have shown that the holdfasts of the larger brown seaweeds do not snap even when pulled by a force of over a ton!

I can see the little bay where I anchored my boat from a headland not far from the old house and often watch as storm-driven waves beat against the rocks with thunderous noise and terrifying power, hurling white spumes of spray in all directions. Although seaweeds belong to the algae, the most primitive class in the plant kingdom, many live for a long time and take many a fearsome beating, so they have great need of their holdfasts. If you stroll along the strand line after a storm you will find it littered with dying seaweeds which have lost the battle against the elements, but those still twisting and turning with the ebb and flow of the tide are the winners.

Beneath the mighty brown weeds I found the more delicate fronds of the red algae, looking like tiny trees, their branches intertwined to form a colourful underwater forest. Starfish, crabs, shrimps and prawns, as well as small fish, moved in and out of the 'trees' and the illusion of a forest was completed by a variety of green algae, each providing its own unique shape and shade of colour. Some species are not useful enough, or perhaps not common enough, to have been given everyday names, but my underwater visit had, for once, a set purpose – to gather as many algae as I could which had at one time or another been useful in local, national or even international enterprises. I am not a particularly good diver (my great-grandmother gathered almost as many just by wading) and it was with much puffing, blowing, snorting, grunting and coughing up of seawater that I carried the weeds to the surface. Eventually my buckets on board the boat were filled with a slimy assortment of greens, reds and browns. Once out of the water their beauty collapsed, but a seaweed collection can always be revived in a deep tray of seawater. My catch complete, with the *Little Witch*'s outboard chugging away, I headed for a deserted stretch of beach where, on the previous day, I had already laid a driftwood fire and hidden a box of provisions. Reaching into my bucket, I soon had a fine collection spread out on the sand next to a crackling fire. All was now set for a journey back through the centuries to re-enact the early days of the alkali, iodine and food industries which evolved on the sea-shore.

The Alkali Industry

Before the discovery of a chemical method of making soda directly from salt 'natural alkali' had to be obtained for the essential and, after the seventeenth century, rapidly expanding industries of soap and glass. Soap was discovered early in our civilisation, but its manufacture was a long-drawn-out and expensive process. It was once boastfully proclaimed that Queen Elizabeth I took a bath with soap twice a year, whether she needed it or not. This makes you wonder what the hygiene of the rest of the population was like! Soap is made by reacting fat from an animal or plant with an alkali. The fat contains stearic acid and this combines with the alkali to produce soap, which is a neutral salt, with water as a by-product. Glass was in use around 6,000 years ago and was made by reacting silica (the geologist's name for sand) with alkali.

The main problem with both processes was the provision of a regular supply of alkali, and as usual man first turned to plants.

Natural alkalis were first produced by burning trees or other vegetable matter in pits, shaking up the ashes with water and evaporating the resulting solution in large iron pots in order to concentrate the chemicals absorbed from the soil by the plants. Hence the name potash.

By Tudor times timber supplies were already dwindling alarmingly and from the eighteenth century onwards large amounts of potash were imported from both North America and Russia, with consequent damage to their forests, until transport costs became prohibitively high. Attention then turned towards alternative sources of ash and it was found that seaweeds were an ideal substitute, although the end product was not identical. Seawater is rich in sodium, but is comparatively poor in potassium. This does not matter much, since these two elements are closely related and have similar properties. It was found that soda was an adequate and much cheaper substitute for potash and could be prepared easily by burning seaweeds, which were collectively known as kelp. Yet the seaweed industry in Britain did not develop quickly since British glass-makers preferred to import barilla from Spain. This was produced by burning the seashore plant appropriately called *Salsola soda*. But the folly of importing soda from afar became obvious during the wars of the seventeenth and eighteenth centuries, and so the British kelp industry was born.

The Kelp Industry

Any rocky coast around Britain is rich in brown seaweeds and with a little practice it is possible to identify the various species. The best species for kelp production belong to the *Laminaria* and *Fucus* families. They are also of interest from the botanical point of view, since different species are found growing in different zones of the shore.

The zonation of the brown seaweeds can be clearly seen if you follow the ebbing tide until the low-water mark is reached. In this zone and also in the water are found the *Laminaria* family of seaweeds and their relatives, which are often referred to as oarweeds.

Tangleweed *Laminaria digitata*

This owes its name to the long finger-like projections on the frond. Those fronds with 'fingers' seem better able to break up the strength of a wave than those without fingers. This species is therefore able to live in very exposed conditions and there is a suggestion that *Laminaria digitata* develops more fingers when growing in exposed situations than when growing in a sheltered bay. It is common throughout Europe and also along the Atlantic coastline of North America, where it is known as devil's apron. In Britain sea tangle, sea girdle and tangleweed are names frequently used. It can grow up to 3 metres long and is often over 1 metre wide. Its basic colour is brown, but when struck by sunlight it shows an attractive golden flush. Because of its size and weight, tangleweed was popular with the kelp harvesters, who required 10 tons of wet weed to produce 1 ton of hard black soda ash.

Sweet Wrack *Laminaria saccharina*

This interesting species has sometimes been called 'the poor man's barometer'. My great-grandmother was convinced of its usefulness, and with it produced a local weather forecast that was surprisingly accurate. She called it 'sea belt' and nailed a piece just inside the garden shed. She predicted wet weather when the frond was swollen and succulent; and an improvement in the weather, when the drying atmosphere drew out water from the weed, which then became dry and brittle. Ten metres long or more and up to 30 cm across, the plant produces a thick, sugary liquid when it is wet, which no doubt accounts for its name of *saccharina* and for a number of its vernacular names, such as sugar kelp, sugar wrack and sweet wrack. As the seaweed dries, white crystals of the sugar (which has been identified as mannitol) can be seen on the surface. This sugar, when extracted, is finding an increasing market in the modern world. For example, it is dusted on to chewing gum and added to pills to mask the unpleasant flavour of medicine. It is also used in the manufacture of top-quality paper, shoe polish, varnish, solder flux, in the leather industry and even in the production of explosives. Research into such seaweeds is increasing fast and previously unknown species are likely to become part of vital industries in the near future. *Laminaria saccharina* may soon be commercially farmed, but for the moment it must be removed from rocks near the low-water mark. It is found throughout Europe and along the Atlantic coast of North America attached to rocks and pebbles close to low water and also in deep rock pools. A study of how it grows in these areas may well be the key to its future production. The frond itself is leathery, has a crinkled margin, a thick smooth round stalk, which also tastes sweet, and a powerful holdfast which grips the rock like a vice.

Oarweed *Laminaria hyperborea*

This differs from the other oarweeds in having a stiff, erect stalk (called the stipe), which is much less flexible than those of

Sweet wrack (**left**), dabberlocks (**centre**) and
tangleweed (**right**)

the rest of the family, so that it tends to stick out of the water at low tide, making it easily identifiable. It is common on the lower shore and in rock pools around Britain and Norway. The frond is up to 2 metres long, but at 1 metre it is comparatively broad. It is divided into fingers, but is distinguished from *L. digitata* by its size and shape and by its erect fronds. The holdfast is particularly strong and although the plant may be battered and torn during storms, it can usually hang on and repair any damage during the growing season the following summer.

Dabberlocks *Alaria esculenta*

This plant, usually called dabberlocks or brown ribweed, occurs among the Laminarians, but can be distinguished from them by the conspicuous mid-rib on the frond. It is a very common species throughout Europe, and on North America's Atlantic seaboard, where it is called winged kelp. In Scotland it used to be called henware, although I have not been able to discover why. An unusual feature of dabberlocks is the reproductive structures branching off from the stipe close to the holdfast. These stick out like little leaves and may well be given some protection by the more powerful frond. The male and female cells join to produce an embryo which is carried by the sea until it finds a suitable habitat where it can grow. This method is thought to be an improvement on those species which have their reproductive organs on the frond itself and so are subjected to the full force of the weather.

After the zone dominated by the Laminarians we come to the *Fucus* family and their relatives, which are also brown seaweeds. They all contain the green pigment chlorophyll, which helps speed up the process of photosynthesis by which the plants make food in the form of sugar from carbon dioxide, water and sunlight. In brown seaweeds the presence of chlorophyll is masked by a brown pigment called fucin, which is tough and may give the plant a measure of protection from the force of the sea. Starting from the *Laminaria* zone and moving landwards, the following species occur, each dominating its own zone. Serrated wrack is followed by bladder and knotted wrack, which tend to be co-dominant, then comes flat wrack and finally channelled wrack.

Serrated Wrack *Fucus serratus*

This species is also known as toothed wrack, because its flat fronds have saw-like edges which serve to break up the force of the waves, and as black wrack. The fronds often appear to be covered with blisters. These are, in fact, the reproductive organs, those of the males

Serrated wrack

Knotted wrack (**left**) and bladder wrack (**right**)

being a much brighter orange than those of the female bodies. Serrated wrack quickly dies when dry and this explains why it is always found on the lower part of the shore. It is found commonly throughout Europe, and in Scotland it was one of the most important species collected for the production of soda.

Bladder Wrack *Fucus vesiculosus*
Knotted Wrack *Ascophyllum nodosum*

Bladder wrack used to be known as kelpware and seawrack and is one of the most common and easily recognised seaweeds occurring throughout Europe and along the Atlantic coast of North America.

It is dull green in colour and its fronds end as two short, blunt branches. There is also a very prominent mid-rib and pairs of bladders which are oval and filled with air. These work like water-wings and when the tide is in allow the plant to float near the surface, where there is plenty of light. The holdfast secures the seaweed to the rock and there is no need for roots, since water and salts (particularly sodium) are all around the frond in solution and can be easily absorbed. *Ascophyllum nodosum* (knotted, knobbed or egg wrack) shares the same habitat as bladder wrack and can be distinguished by its single air bladders, which are egg-shaped, and by the special reproductive structures that branch from the side of the narrow frond and look like small, wrinkled grapes on stalks. These greenish-coloured growths are often found in pairs, especially during April, when reproductive activity is at its peak.

I well remember days spent with the little witch gathering knotted wrack to spread on the garden as fertiliser. She relieved the boredom by selecting a particularly large egg, cutting it away from the frond and making a hole in it to pro-duce the most wonderful whistle I ever heard. In parts of Cornwall knotted wrack is called sea whistle.

Flat Wrack *Fucus spiralis*

Flat wrack is found quite near to land and spends a considerable time out of water, often exposed to strong sunlight. If it is picked up and held dangling, its frond is seen to twist, which accounts for its North American name of spiral wrack. Flat wrack, however, is a good name because this species lacks bladders on the frond, which has a smooth edge and ends in a forked tip. The mid-rib is prominent and the colour is olive-brown, but when dry it darkens almost to black. Each frond can measure up to 40 cm but despite its comparatively small bulk it was popular with kelp hunters, probably because it is covered by the tide for a shorter period than the species further down the shore. During and just after the Second World War the little witch and I often helped a 'young man' of sixty-seven to load his cart with 'twisted wrack', as he called it. The weed was crammed tight into wet, stinking sacks and carted off to the railway station. Rumour had it that 'the government used it'. It was probably used for the extraction of iodine for first aid kits.

In all these zones one species predominates, but in transitional zones two species may exist together. This is certainly the case with flat wrack and channelled wrack, the species found closer to dry land than any of the others.

Channelled Wrack *Pelvetia canaliculata*

Because it is so close to land channelled wrack has been used in Scotland as cattle food and has been called cow tang. I have also seen sheep eating it and on one memorable morning I watched red deer enjoy-

Flat wrack (**left**) and channelled wrack (**right**)

ing a breakfast of *Pelvetia* on the shore of the island of Scarba off the west coast of Scotland. My boat was bobbing at anchor as I took an early breakfast on deck under a clear blue sky typical of a May morning in Scotland. The nine deer trotted down to the sea from the cover of a belt of gnarled old birches. Their hooves clattered on the pebbles as they splashed about in the shallows and began to tear at the wrack. A grey seal looked balefully down at them from its rock before sliding gently into the ebbing tide, then headed out to sea in search of its own breakfast.

The fronds of channelled wrack are up to 15 cm long and very narrow. The dark green colour gives way to a shiny black when dry and the frond turns inwards to create a channel. This reduces the surface area exposed to the air and cuts down water loss. Each plant is fastened to a solid object by a disc-like holdfast of surprising strength, making it hard to remove. Channelled wrack is not bulky enough or common enough ever to have attracted the kelp industries of old, but it has been used for centuries as a fertiliser for fields and gardens. The old lady never returned from her wanderings on the beach without a handful of 'tang', which she placed on the compost heap in the garden. Her allotment seemed always to produce better crops than any of our neighbours'. Perhaps it was luck, but I rather think it was 'green cunning' at work again.

While it has never been an economic proposition, channelled wrack, like all the brown seaweeds, can be used to produce soda.

Soda Production

No technique is simpler than the production of soda from kelp and there is no basic difference between my bit of fun on the beach and the technique used when the kelp industry in Scotland was in full swing and employed over 100,000 people. This employment revived the fortunes of the Western Isles because it encouraged the people to return after the harsh and unjust treatment they had received at the time of the highland clearances following the failure of the 1745 Jacobite rebellion.

The flames of my roaring fire soon gave way to palls of smoke as wet seaweed dried, bubbled with steam, cracked and twisted in the heat, then caught fire. Poked with a long stick, the pile of soda-rich ash gradually increased and I added more weed to the flames. The soda produced from kelp is rich in magnesium, and was found to produce high quality glass. The soap makers were also delighted because salt had always been added to the soap to help it coagulate. Salt was heavily taxed and the fact that kelp soda was already rich in salt reduced their costs substantially.

After the fire had burned for four hours, I let it die down and scraped away the ash. A shaft of sunlight flashed from a tiny globule resting in the hot sand. Here was glass – made by the combination of sand, soda and heat. Not a very impressive piece of glass, it is true, but no doubt a similar observation thousands of years ago was the beginning of the glass industry.

The Decline of the Kelp Industry

Gathering kelp, transporting it to a burning area and gathering up the soda is a laborious operation, since more than 10 tons of wet, heavy weed is needed to produce a single ton of soda. It was therefore simply a matter of time before a synthetic method of producing soda was developed, and it was France that had the greatest incentive to do this. Before the Napoleonic wars France was dependent upon Scottish kelp or Spanish barilla for her supply of soda, and when these were abruptly cut off a quick solution had to be found. Nicolas Leblanc's process was the first answer, but because it produced so much pollution (mainly in the form of hydrochloric acid) it was eventually replaced by the Solvay process. The details of these processes are outside the scope of this book, but they succeeded in destroying the kelp industry totally and the collection of seaweed would have ceased altogether if yet another vital chemical, iodine, had not been discovered in the ash.

Iodine from Kelp

Iodine is a vital element in animal growth. In mammals it is utilised by the thyroid gland to produce the hormone thyroxine, which controls growth. If there is a shortage of iodine, the thyroid gland increases in size in an effort to increase production. This causes a swelling in the neck called goitre. Usually iodine is present in our drinking water, but in parts of Derbyshire it is deficient and 'Derbyshire neck' was once a familiar complaint in the area. At one time the villagers of Padiham, near Burnley in Lancashire, were called 'thick necks' or 'leather necks' for the same reason. Iodine is now added to salt, and goitres are very much a thing of the past.

Seaweeds were once exploited as a source of iodine. The element is present in seawater in small, but significant proportions, and the seaweeds filter it and concentrate it. Some of the large oarweeds accumulate massive concentrations, but even the comparatively inefficient egg wrack can contain over two hundred times the percentage of iodine found in seawater. Although nowadays iodine is produced from inorganic compounds under carefully controlled conditions, iodine in seaweeds remains as a reserve should it ever be needed. The kelp and iodine industries received a boost in the two world wars, but now only a few localised concerns exist and these industries are no longer nationally important. Seaweed, however, has once more made a comeback and is at present one of the world's most valuable self-perpetuating commodities. Indeed, our consumer age would be virtually impossible without it and supplies of some species are now so vital that a great deal of research is being directed to developing potential farming techniques.

Modern Uses of Seaweeds

With my pile of seaweed and my driftwood fire I repeated a process which has played an increasingly important part in Chinese and Japanese life since the seventeenth century. I remember my great-grandmother taking an old roasting tin down to the beach, supporting it on pebbles and kindling a driftwood fire beneath it. She selected a number of red seaweeds and dropped them into a small amount of bubbling fresh water in the tin. After a while the thallus of the seaweed dissolved into a jelly-like mass which cooled and solidified when removed from the fire. She always added a small amount of this material to her jam, insisting that it helped it to set. She also mixed it with herbs to produce ointments which spread easily over bruises, burns and cuts. On both these counts the little witch's 'green cunning' anticipated modern technology for it has recently been proved that agar jelly is both an astringent and a mild antiseptic.

Agar-agar is a collective name, deriving

from the Malayan word for a red seaweed. However, different species of seaweed produce jellies with different properties. This has resulted in some terminological confusion and at the present time the terms used reflect the arguments raging among scientists. The traditional use of agar-agar in western science has been as a sterile medium for growing bacteria, viruses and fungi for experimental purposes. Biologists now generally refer to the jelly-like extracts as phycocolloids, deriving from the Greek for seaweed (*phykos*) and glue (*kolloides*). There are three basic types (although each type varies according to the species it is prepared from), namely algins, carrageenans and agars.

Algins

In 1884, E. C. Stanford, a British chemist fascinated by seaweeds, discovered that extracts from the *Laminaria* and *Fucus* species could produce gelatinous substances named colloids with useful emulsifying powers if reacted with alkali. He found no financial backers on this side of the Atlantic, but forty years later his research was put into practice by an American company named Kelco. They used a species called *Macrocystis pyrifera*, but knotted wrack also has a high algin content. This has been used to stabilise and emulsify a large number of substances, including yoghurt, ice cream, milk powder, custard, salad cream, jam, marmalade, pickles, biscuit mixes, cake, icing, syrup in tinned fruit, sterilised milk, cream and cheese. Derivatives of algins are used to thicken soup and sauce, and to control excess foaming during the brewing of beer. An increasingly important application of alginates is in the transportation and storage of perishable goods, such as fish, fruit and vegetables, and they can even be used to restore an acceptable shape and feel to products which are perfectly edible but wrinkled. By now the reader will have realised that we all eat vast quantities of seaweed. We also rely heavily on algins whenever we wish to look presentable before going out on the town. Shampoo, soap, shaving cream, bath oil, bubble baths, toothpaste and lipstick all owe their smooth texture to algins. If we wish to have a permanent record of our evening out, we take a photograph on film coated with algin derivatives. Our shoes gleam thanks to an algin shoe polish. If we order a cocktail, it comes complete with a cherry which is 90 per cent seaweed. On returning home we may feel like a cup of instant coffee containing a little algin, then need to freshen the kitchen with an air freshener also made from alginates. And so to bed, before facing another busy day, in which perhaps some home decorating has to be done.

The wallpaper pattern is fixed with an alginate, which is also a vital ingredient in the size and paste needed to stick it to the wall. The same compound is added to the paint to make it non-drip and smooth-spreading. Hard work may well bring on a headache and once more the manufacturers of algins have their finger in the pharmaceutical pie. Aspirin, indigestion tablets and potions to bring express relief all contain algins. Ointments rub in smoothly and painlessly because of it, dentists line drilled-out cavities with it, doctors stitch wounds with algin-based surgical thread, and operating theatre staff breathe through sterile masks partly composed of it. The uses are increasing all the time. As a result, the seaweed industry now enjoys a more secure base than at any time in its history.

Carrageenan

Carrageenan includes several species of

Chondrus crispus or Irish moss (**top**) and
Gigartina stellata (**bottom**)

red seaweed, among them the edible Irish moss common throughout Britain and the Atlantic coast of North America. Once again the use of this plant originated in Britain and Ireland, but was commercialised in America. From it are produced a complex group of mucilages and gums which chemists have not yet succeeded in copying in the laboratory. As with the algins, the list of uses is long and getting longer. Carrageenan is the main constituent in the production of slimming food, since it contains a pleasant-tasting sugar which is not attacked by human digestive enzymes. It is also added to toothpaste to prevent it going hard if you forget to put the top back on. It prevents the constituents of mayonnaise from separating, and the manufacture of polishes and stains is a profitable industry depending upon regular supplies of carrageenan. Carrageenan was once only available from China and Japan, but wars and politics have ensured that the United States and the Soviet Union have both invested capital in seaweed research, and in Britain surveys are now being undertaken to map the distribution of seaweeds.

Agars

Different agars are produced from different species, usually red seaweeds. The industry once again has its origins in the Far East, especially Japan. The Second World War stimulated urgent research among Japan's former customers and varieties of agar are now produced throughout the world. The first to realise the importance of the substance, about 1650, were the Chinese, for whom seaweed jelly is still an important food, but the Japanese had copied the process by 1662. The legend about the discovery of agar is as follows. During a journey some of the Emperor of China's jelly was left outside and was found to have frozen. As it thawed it turned into a substance with a paper-like consistency. This substance reverted to jelly when water was added. Here was an early example of convenience food, and agar is still stored in its dry form. The method of preparation is complex these days in order to ensure that the product is sterile and can be used to feed and culture bacteria for medical research. It often involves the addition of sulphuric acid to counteract the alkaline nature of sodium salts. The agar is then frozen into bars, and a great number of these are exported and used, particularly in Holland, in the brewing of beer.

During the Second World War it was found that satisfactory agar could be produced in Britain from the seaweeds *Chondrus crispus* and *Gigartina stellata*, both of which are common enough to make harvesting a worthwhile proposition. The former species is also the subject of experiments in America and Canada. *Chondrus crispus*, also known as Irish moss, carrageen and Dorset moss, is a common red seaweed found growing in rock pools and in the mid-shore region both in Europe and the Atlantic coast of North America. It is usually a dull red colour, but its appearance varies according to the strength of sunlight and also whether the tide is in or out. It glows with an iridescent greenish colour when viewed under water. The tough frond is carried on a short stalk, no longer than 15 cm and usually less than 2½ cm wide. This species often occurs together with the closely related *Gigartina stellata*, which has its fronds rolled inwards – a feature never seen in Irish moss.

Bacteriological agar has been in constant and increasing demand ever since Fanni Herse suggested it to her husband in 1881 and Robert Koch also began to use it for his cultures in 1882. No microbiological research laboratory can function without agar and scientists are nowhere near to discovering an inorganic substitute. Other uses for agar, however, are being discovered at an exciting rate. It prevents many fish products, including tuna, from being damaged in the tin by enveloping them in protective jelly rather like frog-spawn, and the presence of agar prevents fish such as herring from making the tins in which they are preserved turn an unsightly black.

Agar is also used as a sizing material for fabrics, the finest quality being applied to silk in order to maintain the sleek texture. Poorer quality agar is quite suitable for such materials as muslin, tulle and nansook. It is used to give a finishing polish to paper and leather, too. Another important quality of agars is that they are able to resist high temperatures and so food industries can use them to coat confectionery, such as doughnuts, and keep the interior fresh. Waxy agar paper is used as a non-stick wrapping for sticky products. In the manufacture of beers and wines agars provide an excellent clearing agent. In the manufacture of electric light bulbs agar is used as a lubricant to make it easier to draw out the thin tungsten wire used for filaments.

I once spent an hour discussing the economic value of seaweeds with a class of fifteen-year-old girls. They were happy to put on lipstick, chew gum, clean their teeth, use sun tan lotion, wear silk, put iodised salt on their chips, eat the agar jelly surrounding the meat in a pie and finish off with jelly (made from agar gelatin) and custard, plus a glass of clear wine. But they all gasped in horror when I suggested that they might like to eat some seaweed.

Seaweed as Food

'They be as good for 'ee as greens,' asserted the little witch as she persuaded me to chew the rubbery-looking stalks of sea tangle and bladderwrack which she had rinsed under the tap and squeezed dry in a cloth. She insisted to her dying day that this was good for children's teeth. I had my first small filling when I was twenty-three and have had five others since. She also insisted that dried seaweed added to a bread mix made it good for invalids. It is now known that seaweed carbohydrate does not increase blood sugar levels and is non-fattening. Following work done in Germany between the wars by Dr Heinrich Linau, several European countries are experimenting with the idea of producing flour made from Laminarians in the hope that patients suffering from high blood pressure and diabetes may be safely nourished.

There is far too much resistance in this country for us ever to consume seaweeds on the same scale as China and Japan and it may be that it is something of an acquired taste. Thanks to the old lady, I developed the taste early and still enjoy a meal of laver bread and sea lettuce, with a few additions provided by a little 'animal cunning'.

Purple Laver *Porphyra umbilicalis*
The little witch always called this plant black butter or laver bread. It is found on the rocks throughout the shore. The fronds are of very irregular shape, looking like wet lettuce leaves, though a great deal darker, and if you rub them between your fingers you will notice two thin layers of tissue. In the early stages the 25 cm diameter fronds are green, but soon darken to purple and then to 'black' when they are considered ripe. The species is in fact one of the red algae.

Purple laver, also known as laver bread

Perhaps black laver may be a better name than the more usual purple laver. It is common and is found throughout Europe and North America. It can look most unappetising when it is left high, dry, black and shrivelled on the beach, but it soon recovers when washed in fresh water and left to soak.

I had left some laver bread in a bowl of water for most of the day, pausing now and again to stir it with a piece of driftwood scrubbed clean with sand. My seaweed hunting, kelp burning and agar brewing had made me hungry and ready to end the day with a seaweed picnic, so I emptied the laver into a small frying pan and mixed it with an equal amount of oatmeal. I then added a piece of fatty bacon. The smell of this delicious fry-up set my mouth watering as I remembered the day when my great-grandmother found a mallard's nest containing four fresh eggs, took two and mixed them with the laver, so I dropped an egg into the pan and completed the dish with a few cockles, which abound in the sand of my sheltered bay. All that was left was to add the green vegetable and there is no shortage of this on any beach, as sea lettuce has a world-wide distribution.

Sea Lettuce *Ulva lactuca*

Sea lettuce has been used as food for centuries and because it has a similar shape to laver bread it is often called green laver. Young plants are pale green in colour and inexperienced seaweed hunters often think they are old specimens 'going off'. Mature specimens are a delightful fresh green and the fronds, which can be up to 45 cm in diameter, look very appetising. Male and female plants tend to be separate but cannot be distinguished unless viewed under a microscope, when the mobile male cells can be seen. Sea lettuce grows in delicate tufts and the leaves, which do not have a double layer of tissue like purple laver, are commonly found washed up on the shore.

My great-grandmother had several recipes for making sea lettuce into a tasty dish. One was to wash it and then drop it into hot fat for a second or two. It is then served with laver bread. I like it best rinsed, then boiled in water with a dash of salt and a splash of vinegar. After draining through a colander or squeezing gently with a cloth it can be served straight away.

I sat on my lonely beach, enjoying my meal, and as I looked at my soda ash rich

Sea lettuce

in iodine and my bowl of coagulated agar jelly, I wondered what else is contained within seaweeds and awaiting discovery. The tide was beginning to turn, the light was fading and the sand dunes were etched against the red disc of the setting sun. I smiled as I thought of all the exciting plants which abound on these dunes and which had given both the little witch and myself so much pleasure.

4 Plants of the Coast

Seaweeds cannot exist in areas which are not covered for at least part of the time by the tide. Apart from the fact that they have no roots and obtain all their essential salts from the seawater, they also need water to reproduce. The male cells reach the female cells by swimming through the water and the fertilised eggs develop into embryos which are dispersed by the currents.

With one or two exceptions, land-based plants with a root system or with flowers designed to be pollinated by insects or the wind find conditions in the sea impossible. The best way to appreciate the gradual take-over by land plants is to go for a paddle and follow out the ebbing tide. This is quite safe if you never venture above ankle depth, but however well I know the area, I always carry a stick with which to probe the route. When the low tide-mark is reached, if you turn round to face the land, you will notice a wet but drying area populated by a few species of the most adaptable flowering plants in the world. A quick glance will show that they are not sea-weeds, because they have fully functional roots, stems, leaves, and, at the appropriate season, flowers. Two such plants found on British beaches are eel grass and glasswort, both of which have had their uses in the not too distant past.

Eel Grasses *Zostera* spp
I was once cleaning out an old attic which contained a dilapidated chair with its stuffing hanging out. The strands did not look like horsehair and when I examined

them under a microscope there was no doubt that they were plant material. This turned out to be eel grass, the collection of which was an important cottage industry in some coastal districts in England until the 1930s, when a mysterious virus devastated the plant. For a time the industry fought on by importing a species with similar properties known as Algerian grass from the Mediterranean. It was also used as packing material for china and glassware. There are in fact three species of eel grass – which are not grasses despite their name and the grass-like appearance of the leaves, which can be up to a metre long. The smallest species is known as *Zostera nana*. This is

Eel grass

by far the toughest and occurs in the most exposed situations. When covered by water it wriggles about sinuously like an eel, hence its vernacular name. The other two species, *Zostera marina* and *Z. hornemanniana*, are very similar but were reclassified as separate species by T. G. Tutin in 1936. It was found that *Z. marina* does not have notches in its leaves, a feature which is typical of *Z. hornemanniana*. The small green flowers are enclosed in a protective fold and are unusual in being pollinated under water. The peak flowering period is August, but I have found fertile blooms as early as March and as late as mid-October. At one time it looked as if eel grass might become extinct, but either the virus responsible has weakened or *Zostera* has developed a resistant strain. As a result a dramatic recovery is well on the way and birds such as Brent geese and wigeon, which up to the 1930s ate eel grass and very little else, are showing signs of returning to their traditional food. But the packing industry will never return to the beaches, for expanded polystyrene is here to stay.

Glasswort *Salicornia europaea* agg

Few plants cause so much debate among professional biologists as glasswort. Is it one species or several? Is glasswort an annual or a biennial? The compromise solution may be seen in the term 'agg', for here is an aggregate of similar and probably interfertile plants. The little witch did not care and neither do I, as I collect the succulent green stems, which are at their best during August and September. These are plunged into vinegar and can be used as pickles for sandwiches. They taste rather like gherkins.

We used to call the plant samphire, but this is misleading because glasswort is not in any way related to true samphire (*Inula*

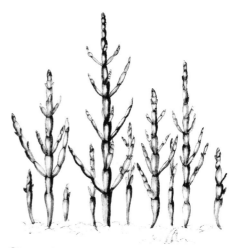

Glasswort

crithmoides). Glasswort is still eaten in some areas, especially along the estuary of the river Wyre near Fleetwood, but its industrial use has long since disappeared. As its name implies, it was burned along with the kelps to produce alkali soda for use in soap-making and glass-making. *Salicornia* has juicy green stems which poke out above the mud to a height of 15 to 30 cm. The stems are much branched and look like tiny succulents or spineless cacti. The leaves, however, have been reduced to minute scales in a successful effort to cut down the loss of fresh water, which is just as vital to a shore plant as to those growing miles inland, with no contact with salt water. The roots are surprisingly shallow and a storm-force tide can easily drag them from their anchorage. The plant is so well adapted to its environment that even such a potential disaster is turned to advantage, since the disturbed roots seem to be stimulated to grow even faster and, if the flowers have been pollinated, some of the seeds are dispersed to new areas. The peak flowering period is early September and, like those of eel grass, the tiny, green flowers are insignificant-looking and are tucked away under a branch. A low-power magnifying glass is needed to pick out the anthers covered by a dusting of yellow pollen. Although it cannot be proved, it is highly likely that glasswort is becoming more common now that its use in soda production has declined. This is probably also true of another plant found growing in the drier areas of sandy beaches, but still within splashing distance of the sea. This is the well-named prickly saltwort, which is a regular member of the strand-line flora.

Saltwort *Salsola kali*

This low-growing plant, as its name implies, was also burned for its soda ash. It is now usually studied closely only by swimmers, when they step painfully on the thorny leaves. The prickly saltwort contains a high percentage of soda, but the plant was never common enough to encourage large-scale harvesting. It was used in glass-making, however, as far back as Old Testament times. Prickly saltwort is an annual plant and its erect and prickly leaves seldom exceed 60 cms. The branched stems are usually pale green, but can sometimes be striped a distinct red colour. The tiny flowers are solitary and hidden in the axils of the upper leaves. The thick coating of wax on the leaves is another successful strategy to cut down water loss.

Scurvy Grass *Cochlearia officinalis*

Scurvy grass is another attractive plant of the strand line. During the war when the supply of fresh fruit – particularly citrus fruit – dwindled, some people, especially children, began to show signs of vitamin C deficiency. Dry, chapped skins and bleeding gums were the first signs. But in our family no one suffered from symptoms of this kind, since the little witch persuaded us to eat scurvy grass, despite

Scurvy grass

its bitter taste. Not all her knowledge was learned at her mother's knee, however, and she had well thumbed copies of all the important herbals. I remember her reading Gerard's account of the horrors of scurvy and eating up my *Cochlearia* without further argument. I have her copy open in front of me now and the old master botanist certainly knew scurvy well. He describes it as 'a filthie, loathsome, heavie and dull disease'. He also knew the symptoms: 'the gumms are loosed, swolaen and exulcerate; the mouth grievously sickening; the thighs and legs withall very often full of blewe spots and the feet swolne as in dropsy.'

The fact that scurvy grass was a reliable cure was well known in Elizabethan

England and it was often sold on and around the docks. The vendors had their own street cry:

Hay'ny wood to cleave,
Will you buy any scurvy grass?

The bitter taste was often masked with additives, but no sailor would object to chewing it when he had seen his mates die horrific and painful deaths. It is known that Captain Cook's ships all had a supply of scurvy grass put into their water barrels before setting off for the South Seas. It would not last any great length of time, however, and the plant went out of favour when the British navy discovered the value of citrus fruits, particularly limes. These last for a long time and their use on British ships in the eighteenth century was the reason why the Americans, still chewing their scurvy grass, called us 'limeys'.

I have always found scurvy grass palatable when it has been soaked for a while in fresh water, chopped and added to a salad. My great-grandmother served it with a mussel or flat-fish stew. Its use in this manner is not surprising because it belongs to the Cruciferae family, which also includes watercress, cabbage, Brussels sprouts, swedes and other important vegetables. The cultivation of most of these developed only during the agricultural revolution of the eighteenth century. The crucifers all have four petals arranged in the form of a cross, those of scurvy grass being white, although they are occasionally tipped with purple. The leaves of scurvy grass are a rich green and very succulent, typical of a plant growing near the sea, since for seaside plants an efficient system of storing water is absolutely vital. Another notable feature is the seed capsules which float, so that spring high tides will carry some of the seeds away and allow new colonies of scurvy grass to develop. Although it seldom grows higher than 30 cm, scurvy grass dominates many seaside areas, especially land sloping down to the sea. Scurvy grass can be found in flower from March to October, but this is because four species occur in Britain with different flowering periods. Stalked scurvy grass (*Cochlearia danica*) blooms from March to August; common scurvy grass (*C. officinalis*) from April to July; long-leaved scurvy grass (*C. anglica*) usually only during May and June; and Scottish scurvy grass from June to September.

When common scurvy grass is found growing on the strand line the individual plants tend to be smaller than those growing in a more fertile habitat.

Sea Rocket *Cakile maritima*
Sea rocket is another lovely cruciferous plant of the strand line. The branching stems are often around 30 cm tall and the fleshy, lobed leaves are dark green in colour. The plant does contain vitamin C, but its taste is extremely bitter and is even more unpleasant to the palate than scurvy grass. It is an annual plant, with a very long tap root which holds it firmly in the sand along the strand line. It also holds its place on exposed shingle beaches, where the wind removes many potential competitors. The four lilac or white petals measure 1 cm and are twice as long as the sepals. Sea rocket has a long flowering period and is generally found from June until October, though I once came across a plant in full bloom in Cumbria on Boxing Day.

Species such as the eel grasses, glasswort, saltwort, scurvy grass and sea rocket found in the exposed areas of a beach can be considered as pioneers, the plants themselves and their decaying bodies

Sea rocket

acting as sand traps and producing tiny sand dunes which grow gradually larger. These dunes are covered with colourful and useful plants. Marram grass usually dominates, but scattered among it and protected by it are such species as rest harrow, sea holly, sea bindweed, henbane, burnet rose, dewberry, silverweed, viper's bugloss, lady's bedstraw, dyer's greenweed, toadflax and, perhaps the most interesting species of them all, the evening primrose.

Marram Grass *Psamma arenaria*
Writing in 1939, Dorothy Hartley de-

scribed a hassock made from marram found in an Irish church:

> I reckoned that specimen of grass must have been at least 80, and more probably 120 years old. It was woven in a circular form and had been filled with peat moss, or some similar substance and sewn up to form a hassock or stool. By its position in the depths of the ruined crypt, under stones, dust, and old bones, it had probably been used as a basket to carry down rubbish when the place was cleared. It must have been damped and dried repeatedly and then left forgotten, yet the grass was as elastic and firm as the grass I had seen growing on the dunes at Newborough across the water. Under such conditions straw, reed or rush must have perished, but the marram survived.

Such strength is essential if marram grass is to stand firm on windswept dunes, its long tough roots binding the sand. Although the grass cannot survive being inundated by seawater, it reacts speedily to being buried by sand, growing very quickly until it reaches the light, when growth slows down. This resilient grass was used in many areas, particularly on Anglesey, to make brooms and for thatching houses. I have often wondered why more thatchers did not use marram, since it adapts so well to weather changes. Its leaves are grooved and the interior is lined with hairs, both features serving to conserve water. If marram is collected on a really hot day the leaves will be seen to be curled inwards, then uncurl when placed in water.

Rest Harrow *Ononis repens*
I knew and loved this plant before I could read or write, and it has fascinated me ever since. It was, as usual, all the doing

of the little witch. We had waited all day for the July thunderstorm to abate and a bucket half full of water and a lemonade bottle almost full of water had been ready since breakfast. Just after six, the clouds parted and the sun came out. In a dozen strides we were on the dunes, which were draining quickly. The flowers were sparkling, as sunlight reflected from their dripping foliage. Bees were buzzing and I can remember as if it were yesterday the heady perfume streaming off the beds of thyme and lady's bedstraw. Within five minutes we were kneeling in front of a sandy bank flushed pink with rest harrow. Our small garden forks soon unearthed the straggling roots of the flower, which we cut into strips with an old pair of scissors. My great-grandmother washed the cut roots carefully, pushed them into the neck of the bottle, then stoppered it. I watched in wonder as she shook the concoction, which developed a froth, its colour darkening to a deep brown. I plucked up courage and took first a sip, then a good long drink of the liquid I now know so well – Spanish water. It tastes very much like liquorice and indeed the two plants are closely related. True liquorice is native only in southern Europe, but Spanish water makes a pleasant substitute.

The scientific name *Ononis* comes from the latin word for an ass, and is interesting because it is thought that the spines which often cover the plant deter feeding animals. Some botanists believe that rest harrow growing in poor ground has more spines than where the soil is richer. Linnaeus suggested this in the eighteenth century and he drew a human parallel by pointing out that the richer the habitat in which it lived the more 'tamed' or civilised the plant became. The old English name seems to have been wrest or arrest harrow, whilst in France it is called *arrête-boeuf*. The meaning behind both these names is clear enough – rest harrow's extensive and tough root system makes the cultivation of the rough ground on which it grows difficult, arresting harrows and stopping even ox-drawn ploughs.

Rest harrow is a delight to see and graces sand dunes, and sometimes disturbed areas further inland, with its delicate pink flowers from as early as May to as late as November. It is a low, creeping shrub carrying stalkless leaves and flowers, the whole plant being covered with a fine layer of pale delicate hairs. These are easily seen if you get down on your knees and view the plant against the light. The hairs retain water – a useful feature in any plant if it is to thrive in an area which drains as quickly as a sand dune. The margins of the leaves are notched, another useful water-retaining structure.

A quick look at the structure of the flower shows that it is a member of the Leguminosae, the pea family. The calyx is small, being made up of five sepals, but the complex make-up of the flower is due to the five large petals which together are called the corolla. Two of these petals are joined as a pair to form a closed keel and another pair, appropriately called wings, project from the side of the flower. The fifth petal, called the standard, is much larger and darker than the rest and stands upright. There are ten stamens joined together to form a tube, which in most members of the pea family serves to guide insects towards the nectaries. Bees very often pollinate the Leguminosae, but rest harrow does not play at all fair. It provides the tube, but not the nectar. So why do bees continue to visit rest harrow? The German biologist Muller carefully observed bees visiting the plant and pointed out that whilst the drones quickly learned that rest harrow was useless, the workers collected large amounts of pollen, which is essential food for the grubs. He thus established that the plant is pollinated by worker bees. The mechanism involved is fascinating. The keel interlocks with the two wings in such a way that any pressure on one is transmitted to the other. The keel forms a perfect landing platform for any passing bee and its weight causes the keel to give way. The stamens then push out like a piston and the bee is dusted with ripe pollen. When all the pollen has been used, the female stigma becomes ripe. When a bee covered in pollen lands on such a flower, the stigma is thrust out, its sticky surface picks up pollen grains and cross-pollination is thus achieved. The ripe seeds are carried in pods, technically known as legumes, from which the scientific name of the pea family is derived.

Sea holly

Sea Holly *Eryngium maritimum*

Another of the little witch's treats was a trip to the dunes in August to dig up the roots of the plant she called 'ringo'. This attractive plant is found around the coast of England, Wales and Ireland, but has never been common in Scotland. It suffered from overpicking and is now rare on beaches, where it must once have been common, so it must not be picked or damaged.

The leaves of sea holly are superficially very like holly, but it is a member of the Umbelliferae family, which includes such plants as parsley and carrot. The flowers vary from blue to purple and are bunched tightly at the top of a thick, rigid stem which can be as tall as 60 cm. Sea holly is edible and at one time the young shoots were prepared and eaten in the same way as asparagus. The roots were known to have a value even in ancient times and the Greek physician Dioscorides recom-

mended 'eringo' as a treatment for flatulence. It was as a sweet, however, that the greatest demands were made on our native stocks, and the little witch had a very tasty recipe. She first scraped the roots and stood them in honey diluted with hot water for twenty-four hours. Then she simmered them gently for a couple of hours until the sugar had soaked into the tissues and left them to dry, the sugar often crystallising and glistening in a most mouth-watering manner. During early August she produced a new flavour by adding wild raspberries, whilst blackberry was the flavour for the month of September. I doubt if the professionals based at Colchester in Essex ever produced better than those I sucked in Cumbria during the 1940s. The trade finally disappeared in the nineteenth century, but the Elizabethans knew eringoes well; in *The Merry Wives of Windsor* Shakespeare has Sir John Falstaff tell Mistress Ford to 'Let the sky rain potatoes; let it thunder to the tune of

"Green Sleeves", hail kissing-comfits and snow eringoes'. Mistress Ford would have had to keep even more of an eye on the lecherous Sir John when he had been eating eringoes if John Gerard's *Herbal or General Historie of Plants*, published in 1597, is to be believed. He grew the plant in his physic garden and hinted at its use as an aphrodisiac good for 'nourishing and restoring the aged, and amending the defects of nature in the younger'.

Sea Bindweed *Calystegia soldanella*

The pink bell-like flowers of this creeping perennial make it one of the most colourful plants found on sand dunes, and occasionally it occurs on shingle. The slender underground stems (rhizomes) anchor the plant and are also important in stabilising the dune. Although common in England, it becomes increasingly rare in Scotland and in the Western Isles grows only on Eriskay, where it is known as 'the Prince's flowers'. Legend has it that the Young Pretender brought the seeds with him from France at the beginning of his 1745 campaign. However, I think the true explanation is scientific rather than historical. The climate of Eriskay is better than many nearby islands and the chemistry of its sand seems to me closer to that of my native Cumbria than to the Western Isles.

Sea bindweed has no medical use, but my great-grandmother used it as an excellent substitute for garden twine and it could well be used by other gardeners today, who often dip deep into their pockets to buy 'biodegradable string' – which is precisely what bindweed is.

Henbane *Hyoscyamus niger*

The little witch knew her plants, both good and evil. She told me of the old uses for henbane, but insisted that I never touch it. 'It looks evil,' she warned, 'an'

Henbane

it be evil an' all.'

Henbane is common in Britain and grows well on sand dunes and shingle. I have, however, frequently found it growing inland, usually in the vicinity of old village churches and especially monasteries – a sure sign of its value as a medicinal plant. In fact, it was one of the main plants grown in apothecaries' gardens. It is a sticky, hairy plant and reaches a height of between 30 cm and 1 metre. Large numbers of funnel-shaped flowers are arranged in rows all along one side of the stem and, while some flowers have short stalks, some are joined directly to the main stem. The corolla is of a creamy yellow colour, covered with purple veins leading towards a dark centre. I have heard henbane flowers described as a 'pale face covered with varicose veins'. The leaves are large and serrated, the upper ones clasped tightly round the stem. When the flowers have died down, a globular capsule develops which contains many tiny seeds. These are shaped like a jaw-bone and henbane was used as a signature plant in the treatment of toothache. Despite the unpleasant smell, the capsules were smoked by country folk suffering from toothache and the danger of this practice was increased by the fact that it sometimes worked rather like codeine does today. There were also alarming side effects, including convulsions, and Gerard was well aware that henbane could induce 'an unquiet sleepe like unto the sleepe of drunkeness which continueth long, and is deadly to the party'. The chemicals responsible have been identified as hyoscine, atropine and hyoscyamine, which causes delirium and increased heartbeat, as well as convulsions, and finally death. In an age when forensic science was almost unknown, death by an overdose of henbane was difficult to detect and it was the means chosen by the infamous Dr Crippen to do away with his wife in 1910.

The monks certainly knew henbane and used it to relieve other forms of pain besides toothache. It was simply a question of calculating the correct dose, a task which then as now required training and experience. In modern medicine henbane has been used to dilate the pupil of the eye prior to examination or surgery, and also in the production of 'brain sedatives' and in pills to prevent seasickness. At one time henbane was a witches' plant and when burned on a fire it gave off fumes thought to confer clairvoyant powers. It was also said that bathing the feet in water containing henbane leaves cured insomnia. Drug addiction is not a recent problem and henbane was used by seventeenth-century drug takers, who called it 'twilight sleep'. Because of the plant's soporific powers, babies were put to sleep by soaking a henbane root in red wine and hanging it round the child's neck on a string of bindweed. This potentially lethal 'dummy' soon had the child fast – sometimes too fast – asleep.

Burnet Rose *Rosa pimpinellifolia*
I have already mentioned my great-grandmother's rose hip syrup, which she usually made from the fruits of the dog rose. As the war dragged on, she turned me into something of a professional botanist. The government had chosen rose hip syrup as a source of vitamin C for wartime infants and contacted every country school with an offer of 3d per pound for rose hips. How well I remember 'Old Reggie', the headmaster, reaching into a blue bag and taking out piles of shining copper threepenny bits, with their angular shape and thrift flowers on the reverse. It was quickly realised that burnet rose hips contained more vitamin C and for these we could get 4d per pound. With a twinkle in her eye, the little witch watched as I scoured the Cumbrian dunes in search of the dark purple fruits. The money was hard earned because, as its botanical name indicates, burnet rose is a great deal more prickly than other roses. It is often called burrow rose or Scots rose, and in Pembrokeshire it used to be known as St David's rose and was the emblem of the see of St David's cathedral.

The creamy white flowers dominate the sandhills during June and July. They are fewer in number than those of the dog rose and always occur singly; the leaves are much smaller, being made up of three or four pairs of leaflets. I have also found burnet rose growing on limestone hills and have made an excellent wine from the hips.

Dewberry *Rubus caesius*
It was a long time before I realised that this plant was not one of the hundreds of varieties of bramble that exist in Britain, but a separate species in its own right. Two facts help recognition. First, the flowers and fruit appear before those of blackberry. Secondly, although the fruit looks like a blackberry that has been lightly dusted with pollen, the drupes (or

segments) are much larger. The dewberry plant itself is much less vigorous than blackberry and creeps along the ground, perhaps obtaining support from other dune plants, particularly marram. It also grows well in the damp hollows between the dunes. The leaflets carried on the stem are always divided into three (whereas blackberry leaves have from three to five divisions), the flowers are always white (never pink) and the sepals are always erect. I still make jam from dewberries using a recipe dating back to the time of the little witch's grandmother, whose husband had spent some time in a Napoleonic jail. Dewberries, honey, a dash of thyme, a few crushed crab apples and agar from seaweed combine beautifully spread on a slice of hot home-baked bread.

Silverweed *Potentilla anserina*

I remember the look on my mother's face one Sunday lunch time in the early 1940s. Aunt Doris arrived without warning and honour demanded potatoes with the sea bass I had caught on a night line made for me by the old lady.

'Come, boy, we'll away and fetch some goose flower,' said the little witch. 'Thee cook the fish,' she smiled at my mother, who was now well used to miracles performed by those with green cunning. In a matter of half an hour we had dug up the silverweed roots and they were boiling on the hob. Aunt Doris pronounced the mashed 'potatoes' excellent and, for my own part, I can vouch for their pleasant taste.

Before the seventeenth century silverweed was an essential item of food, and there is plenty of evidence that it was eaten in prehistoric times. Its pollen grains, which like those of most other plants are almost indestructible, have been found in middens (refuse heaps) and amongst fragments of pottery. Before the introduction of the potato, originally brought from North America in the sixteenth century, silverweed was grown as a root crop. The roots were dug up during the autumn and, after they had been washed and scraped, the white fleshy interior was either boiled or roasted. At the time of the Irish potato famine of the 1840s and also during the two world wars some country folk, including us, made use of it. I still eat it occasionally and look forward to the experience.

Early man was not alone in finding silverweed palatable, if the specific part of the scientific name is anything to go by. *Anserina* means 'of a goose' and this accounts for the vernacular name of 'goose weed' or 'goose flower' which was given to it by the little witch. It is also known by a variety of other names, including wild skirret, 'spring carrot', 'sweetbread' and 'bread and butter' or

Silverweed

'bread and cheese'. The generic name of *Potentilla* is interesting and translates as the 'little powerful one', indicating a medical as well as culinary use. It was supposed to cure 'green wounds, hot fits, wounds of the privey parts, the stone, blood spitting and loose teeth', and walkers were recommended to place a sprig of silverweed inside each shoe to prevent blisters. It was used by those who earned their living by carrying messages and I know of one old postman in the Lake District who used silverweed until the day he retired in 1957.

Silverweed is common throughout Britain, thriving on coastal dunes, waste ground and marshland, where its growth can be surprisingly succulent. It is a perennial with a flowering period peaking during July and August, though it can often be found in flower as late as November. The yellow flowers are easily seen against the mat of straggling green leaves. These are each made up of serrated leaflets, the underside of which is coated with silvery hairs. I have found that the leaves of silverweed plants growing on the sea-shore are hairier than those of inland plants, especially on marshland, where there is no difficulty in conserving water. Indeed, on the sea-shore occasionally even the uppermost surface of the leaves is hairy. It is, I suspect, the silvery hairs of the plant which give it its name – even though that is a less delightful explanation than the one offered by Gerard, who claimed that 'the later Herbalists do call it Argentica because of the silver drops which are seen in the distilled water thereof when it is put into a glass, which you shall easely see rowling and tumbling up and down at the bottom'.

Silver rose is another name by which the plant was once known. I think we can safely assume that the philosopher's stone has not been discovered on a sand dune, but the 'rose' part of the name is accurate, as an examination of the flowers will show. Five sepals make up the calyx and there are also five petals together forming the corolla. The petals are about double the length of the sepals and inside these are a great number of stamens and carpels, an arrangement typical of the rose family. If the leaves are infused in boiling water and the liquid is then strained and cooled, an excellent skin-cleansing lotion is produced.

Viper's Bugloss *Echium vulgare*
This lovely flower is typical of southern England, but also grows well on some northern sand dunes. Its delightful blooms look like a rolling stretch of blue sea, especially when the white blooms of henbane grow among it, adding the impression of white horses driven before a gentle breeze. Because of its spotted stem, viper's bugloss was once thought to be an effective remedy for snake bites, but no extract has ever been produced from it which could possibly have been of any use. Why then do I believe that it worked? Anyone unlucky enough to be bitten by an adder is bound to panic, and the heartbeat increases as the victim leaps about screaming. The poison is therefore quickly pumped round the body. The heartbeat slows down if the victim is aware that very few adder bites are lethal – and the rate would decline further if he opened up the area around the bite to allow blood and venom to escape, rubbed in viper's bugloss and believed that it was an effective antidote. Most doctors would, I think, subscribe to this 'placebo' theory. Furthermore, herbalists have recently used the plant to induce sweating, thus reducing temperature, which might also slow down the effect of venom.

Viper's bugloss

Bugloss means 'ox-tongue', a name derived from the shape and rough texture of the leaves. The plant is a member of the borage family and its stiff, hairy, erect stem is between 30 cm and 60 cm in height. The flowers grow on one side of the stem only, the corolla being shaped like a bell with five lobes at its mouth. The buds are pink or reddish purple, but the mature flowers are bright blue.

Lady's Bedstraw *Galium verum*

This pretty and delicate-looking yellow flower is another species common on sand dunes, but it is also found on roadsides and dry banks during July and August. The stems are rather weak and seldom stand more than 30 cm high, preferring to straggle along the ground. Folklore tells us that this plant was strewn on the bed in which Mary gave birth to the infant Jesus. Before this the flowers were white, but they were turned to gold as a reward for serving as 'Our Lady's bedstraw'. The plant once had many uses, one of which was a pleasant-smelling stuffing for mattresses. It was also chopped up and added to milk so that the acid it contains would curdle the milk for cheese-making. In the Outer Hebrides the crofters used to dig up the roots to extract a red dye, but eventually they were forced to discontinue the practice because of the erosion caused. I have made a lovely dye by simmering bedstraw roots in a small quantity of white wine.

The main use, however, of lady's bedstraw was as a styptic, and it certainly does stop bleeding and help heal wounds. It was also used in the same way as silverweed to ease sore feet.

Dyer's Greenweed *Genista tinctoria*

Like lady's bedstraw, this species was important as a natural dye and is found on sand dunes as well as on disturbed areas inland. It is more common in the south and does not occur at all in northern Scotland and Ireland. Although it can reach 75 cm, dyer's greenweed is not easy to spot unless the small yellow flowers are in bloom. Its old name was 'wuduweaxe' and it was taken to America by early settlers. Greenweed was one of a complex of dyes, its function being to give a yellow colour as a basis for green. It was also added to material dyed blue with woad, which it converted into a rich green. Gloucestershire was a famous cloth county during the eighteenth and nineteenth centuries and greenweed pickers there worked hard for little pay. In 1829 J. L. Knapp reported that some of the women employed in the back-breaking task of pulling up the plant by its roots were inclined to soak the crop with water, since they were paid for it by the hundredweight. Dyer's greenweed had been collected around the Kendal area of Westmorland for centuries and the archers dressed in Kendal Green were feared from the time of Agincourt.

Toadflax *Linaria vulgaris*
Flax *Linum* spp

Used by apothecaries to treat jaundice because of the yellow colour of its flowers, toadflax was once a destructive weed of the flax fields. Before it flowers this attractive-looking plant is difficult to separate from the crop, by which time its strong creeping rootstock makes it difficult to get rid of. Hence the insulting name given to it. Linen made from flax may have reached Britain from Egypt via Phoenician traders. The species of flax from the Middle East found the British climate very much to their liking – understandably, since we have some native species, for example pale flax (*Linum bienne*), which has light blue flowers and often grows with toadflax on sand dunes. Another member of the family, purging flax (*Linum catharticum*), is also common, but its tiny white flowers are somewhat difficult to find among the dune and heath vegetation. Because it is so small it was said that the little people made their linen clothes from it – hence its alternative name, fairy flax. As the name purging flax indicates, in the past the plant was used as a purgative, the crushed stems being boiled in white wine.

Toadflax

However, the effect of the potion was so drastic that it soon went out of fashion. The flowers are a little more gentle in their action, but even so their use is not to be recommended.

The species from which linen is made is cultivated flax (*Linum usitatissimum*). It is often found growing wild on waste ground and in the summer many fields in Ireland, in parts of Lincolnshire and around Winchester are a sea of blue when it blooms. A major crop of the Anglo-Saxons, it was taxed by the Normans and in the sixteenth century failure to plant flax to provide sail cloth for the navy was a punishable offence. When flax is cultivated, the seed is planted in April, care being taken to place the seeds close together so that the plants grow tall and straight. Once they reach a height of just over a metre they are pulled up by the roots and the flax fibres are extracted from the stems. With competition from wool (which was cheaper) and eventually from cotton, the English flax industry fell on hard times but revived during both world wars when canvas tents, cordage, fire hose and medical supplies all required home-grown linen. The Irish linen trade is still in a healthy state, no doubt with its share of toadflax in the fields.

The flowers of toadflax are also abundant on sand dunes and are not unlike those of the snapdragon, which belongs to the same family – the Scrophulariaceae, which also includes figwort and foxglove. The yellow blooms of toadflax, however, have a long hollow tail-like structure on the corolla called a spur. This is full of nectar and it guides the long tongues of bees towards the ovules which they fertilise. Inside these are four, occasionally five, stamens and a notched stigma. They mature at different times and this ensures that cross-pollination takes place.

Evening Primrose *Oenothera biennis*

In July and August many dune systems are illuminated by banks of evening primrose, which arguably is the plant most likely to shake the medical world within the next ten years or so, as it offers a real hope of a cure for multiple sclerosis. It is not, despite its colour, a primrose, but is related to the willow herbs. The 'evening' part of its name is accurate, since the pale yellow flowers open at night. It is not a native plant, but was brought from Virginia in the early days of the seventeenth century as a garden curiosity. As cotton began to be imported, many evening primrose seeds arrived by accident and were dumped with the ballast. This is why Southport dunes are almost smothered in evening primrose flowers which originated among the cotton crop flooding into Liverpool. In the late 1940s Dr J. Riley, a biochemist at Liverpool University, gathered some of the plants from Southport and began to investigate the properties of evening primrose oil, basing his work on papers produced in Germany in 1917 by J. Uger. The Flambeau Ojibire tribe of American Indians had long used evening primrose oil as a cure for skin blemishes and asthma and Riley isolated from it gammalinoleaic acid (GLA), the unique substance which is bringing about such spectacular advances in the treatment of multiple sclerosis. Old herbals regarded the plant as a sort of cure-all, especially for whooping cough, period pains and gastric complaints, as well as copying the Indian uses as a cure for asthma and a soothing balm for wounds. But it is only in the last ten years that evening primrose oil has attracted the medical profession seriously and trials of its effectiveness are now being conducted in many countries, with pharmaceutical companies setting aside huge research budgets. Already the oil extract

Evening primrose

wonder drug on our hands, but it has also been proved to help alcoholics and schizophrenics, and recent work in South Africa has indicated that some cancer cells are effectively destroyed by it. Research is also being undertaken at the moment to see if evening primrose oil can help those who suffer from migraine, excessive menstrual blood loss, anorexia nervosa, Parkinson's disease and a host of other problems, including infertility. It is already beyond dispute that relief is given to those suffering from rheumatoid arthritis and, of course, multiple sclerosis.

What does this plant look like? Clapham, Tutin and Warburg (see Bibliography) recognise several species, but the evening primroses are all biennial herbs, varying in height from 50 to 100 cm. The stems are hairy, erect and rather woody. The usually narrow, toothed leaves are arranged spirally around the stem and measure up to 10 cm in length. The flowers are about 6 cm in diameter and there are four soft petals, which overlap each other and tend to open at night and early in the morning to reveal eight stamens. When evening primrose was first brought to England it was cultivated in kitchen gardens and in the early days rivalled the potato as a food plant, the rootstock tasting rather like parsnips. It has also enjoyed a long period of popularity as an attractive garden plant.

has been found to lower cholesterol levels and blood pressure, as well as acting as an anti-coagulant, thus reducing the risk of coronary heart disease. The easing of premenstrual tension and the treatment of benign breast tumours have brought peace to many women, and hyperactive children have also been treated with it successfully. The effects of evening primrose extract on skin disorders, discovered by the American Indians, is now endorsed by the medical profession. If this was all the product could do we would have a

Evening primrose is one plant which, so far as I am aware, was unknown to the little witch and I have only ever found one growing on our local dunes. But she did not limit her herbal wanderings in search of sea plants to the dunes and spent a great deal of time gathering plants from the salt marshes, shingle beaches and cliff tops. From her I learned the secrets of sea lavender, thrift, yellow horned poppy, sea radish, sea beet, sea kale, and samphire in the easiest way possible – by listening, watching and copying.

Sea Lavender *Limonium vulgare*

Although not so richly perfumed as true lavender, which is unrelated, sea lavender has a similar smell and the scent becomes more pronounced when the dried flowers are mixed with water. In July and August the small, purplish-blue flowers brighten up the green monotony of salt marshes and the sweet smell attracts bees, flies and bettles to the abundant supply of nectar. On one occasion my mother was in bed with a heavy summer chill and the little witch and I made our way out on to the marshes and collected a bunch of twiggy sea lavender stems, the longest of which measured about 40 cm. They were in full bloom and dusted with yellow pollen. A kettle was quickly boiled on the hob and the sea lavender thrust into the bubbling water. The perfume filled the house and I was quite envious of my mother as she inhaled the fumes, denied now to the rest of us by the towel over her head. For a while sea lavender was everywhere – under pillows, in drawers, in wardrobes, in the bathroom – for its scent deters moths and, as we soon discovered, sea lavender added to your bath brings a delightful feeling of luxury. Not surprising perhaps, since the name lavender derives from the Latin *lavare*, which means to wash.

Like sea lavender, true lavender (*Lavendula officinalis*), which is cultivated in Britain, especially by the monks of Caldy Island, off the coast of Wales, deters moths and a lavender pot-pourri makes a most effective moth-repellent. Furthermore, the oil distilled from lavender in copper stills was recommended by Salmon in his eighteenth-century herbal as ideal for treating 'the bitings of serpents, mad dogs and other venemous creatures, being given and applied poulticewise to the part wounded. The spiritous tincture of dried leaves or seeds, if prudently given, cures hysteric fits, though vehement and of long standing.' It was also recommended as a 'cure' for the black death.

Although sea lavender never rose to such heights, it was indicated for the treatment of colic, dysentery and 'fluxes of the blood'. It is also of undoubted value as an inhalant for clearing the sinuses.

Thrift *Armeria maritima*

The only time I ever saw my great-grandmother cry was at the funeral of one of my classmates who had fallen victim to polio at the age of eight. The old lady shuffled forward and dropped a bunch of thrift gathered that wet May morning

Thrift

from the marsh. I never asked her why she did this, but many years later I was reading a book about the symbolism of flowers and there I found it – for sympathy, give thrift. Whatever flowers she gathered on our plant-hunting trips, thrift was always ignored. 'There be a right time for 'ee to pick each plant, dearie,' she said 'an' this bain't it, thanks be to God.' I have never picked thrift save for one withered bloom when she herself was laid to rest.

Yellow Horned Poppy *Glaucium flavum*

A frequent member of the shingle community, the yellow horned poppy is one of Britain's most beautiful plants. It has some medical uses, but was once the subject of one of the most outrageously comical stories of quackery. The Royal Society meeting in 1698 was asked to pronounce judgement on a most unusual and potentially profitable event. 'A certain person' had made a pie using the roots of yellow horned poppy in mistake for sea holly, but found it 'hot'. He then became delirious, but felt better after he had voided into a white chamber pot a stool which he thought might be gold. It is a pity the minutes of the meeting failed to record laughter. No one seemed suddenly to get rich.

The yellow horned poppy is found all around the coast and the old name for it was 'squatmore', squat being the old English word for a bruise. The little witch swore by it and I have certainly used its leaves to good effect as a poultice. Its erect stems often reach a metre or so in height and the yellow flowers, which are found from June to the end of October, are nearly 8 cm across. The petals are very delicate and soon fall off when handled, although they seem surprisingly resistant to wind. A seed case, often as long as 25 cm, develops when the flower has been pollinated and it is this 'horn' which accounts for the name. The large green leaves are full of water, the lower ones being covered with short hairs, perhaps in order to cut down water loss.

Sea Radish *Raphanus maritimus*
Sea Cabbage *Brassica oleracea*

It is difficult to imagine how we would nourish ourselves without use of the Cruciferae family, which includes so many of our garden plants and vegetables.

The pale yellow or, occasionally, white flowers of the sea radish appear around July. This bushy plant grows to a height of up to a metre and can be recognised by its seeds, which are carried in structures that look like a string of beads with constrictions between them. Some writers suggest that it is restricted to the south coast, but I have often found it on the Cumbrian coast.

Sea cabbage, on the other hand, is very much a southern plant. It was the ancestor of our garden cabbage and was gathered particularly in the Dover area, despite its rather bitter taste. The value of cabbage was known to the Romans and it was supposed to be good for deafness, poor eyesight, gout, baldness and ruptures!

Sea Kale *Crambe maritima*

This rather rare species is also a crucifer though the white flowers, which occur in clusters from June to early September, need to be examined closely to prove this. Sea kale thrives on shingle beaches and reacts to being buried by storm debris by growing very rapidly until it reaches the light. There is no doubt that its value as a food crop has led to its rarity. It is one of the few truly English vegetables that does not grow in the Mediterranean and it was considered delicious enough when blanched to merit mention by the famous

Sea beet (**left**) and samphire (**right**)

French chef Marie-Antoine Carême in his book *L'Art de la Cuisine Française*, published in 1834. He recommends serving 'sea colewort' with buttered toast or fried in breadcrumbs, with a sprinkle of lemon juice and with a béchamel sauce.

Sea Beet *Beta maritima*
Although still found around the coast of Britain, as well as throughout Europe, North Africa and Asia Minor, sea beet has been reduced by collection. Long ago in Persian gardens beet was cultivated both for its leaves and for its roots. Centuries of experimentation, selection and breeding have produced mangold, red beetroot and the invaluable sugar beet. Beet was imported to Britain, but was unknown as a wild plant here until 1629. It was left to the Germans in 1747, and later the French, to develop the plant to its present state.

It is still possible to find 'wild beet' on shingle beaches and cliffs, where it seems to thrive in the atmosphere of sea spray. It tends to straggle a little and seldom grows higher than 60 cm, but the large, leathery leaves are typical of shore plants in their succulence. The flowers, which develop in the axils of the narrow bracts, are in-significant in size and, because they are green in colour, are easily overlooked.

Samphire *Inula crithmoides*
This tough plant never grows much more than 30 cm high, as a result of the exposed conditions on the sea cliffs where it grows. Its vernacular name is a corruption of 'St Pierre', the patron saint of fishermen, an allusion to its position near the sea. The flowers are greenish-yellow and are typical of the Umbelliferae family, which also includes hogweed and carrot. The leaves are fleshy and the plant, preserved in vinegar, was a pickle of such repute that it brought prices high enough to persuade many an adventurous lad to risk his neck collecting it. Shakespeare knew only too well the lengths to which samphire gatherers would go, and in *King Lear* he wrote:

How fearful
And dizzy 'tis to cast one's eyes so low!
The crows and choughs that wing the
   midway air
Show scarce so gross as beetles; half way
   down
Hangs one that gathers samphire,
   dreadful trade!

5 Woodland Trees

My great-grandmother was just as much at home in the woods as she was on the sea-shore and the most memorable day we ever spent together was my tenth birthday, in the middle of September. I opened my presents but there was nothing, not even a card, from the little witch, who was conspicuous by her absence. She appeared suddenly just as I had finished my breakfast.

'Come 'ee with me and Oi'll be showin' 'ee summat,' she grinned, and set off without waiting for an answer, carrying with her a basket full of brown paper packages. Despite her years, she moved with eager steps, her straw bonnet catching the rays of Indian-summer sunlight. Over the dunes she went, through the blackberry thicket without pausing to sample her favourite fruit, and I could feel her excitement. Had she found a new badger sett? Was it too late for the fox cubs? Why had she forgotten my birthday? When she reached a clearing in the wood, she sat down in front of an ash sapling.

'That be yours,' she said, grinning until her features were even more wrinkled than usual. 'Be 'ee confused, boy?' I must have looked it, for she continued, 'I planted this ash when 'ee were born – my ma planted yan for me when I be a bairn.' Many years later I read that it was customary centuries ago to plant a tree to celebrate a birth and that the practice possibly dated back to pre-Christian civilisations. If the tree grew strong, the child would be healthy; if it withered, the child would be frail. I must have looked strong enough at ten years old as I stood beside the sturdy sapling.

The old lady held out her basket. 'That be yours an' all, laddy,' she said and I sensed the tear in her eye. The basket was surprisingly heavy and as I removed the brown paper from the parcels a treasure-house of goodies was revealed. There was a small blue paper bag full of silver three-penny pieces all bearing Queen Victoria's head, and beneath these was chocolate. Then came three heavy, well-wrapped parcels, which with trembling fingers she helped me to unwrap. Out came *Wayside and Woodland Trees* and two volumes of *Wayside and Woodland Blossoms* by Edward Step. 'Time to start teaching you some greenwood cunning, dearie,' she grinned, 'an' a grand birthday to 'ee.'

This chapter is being written in the autumn sunlight of my forty-seventh year, sitting in the shade of a splendid ash. I love this ash because it is mine. There is another one in a nearby glade which is almost twenty years old. This one is my son's and he too knows something of 'green cunning'. I could not resist planting one more tree, up on a sandy bluff overlooking the sand dunes, now already gnarled and bent with the salt spray carried on the wind. It is a witch hazel, a species with many uses since its sap is a strong astringent and antiseptic. I think the old lady would have approved of both choice and location.

Woodland Industries

These days even those of us who think we have green cunning squeeze our woodland walks in between buying furniture in the high street and sitting in a comfortable armchair in front of an electric, gas or, if we are very lucky, a real coal fire. Indeed, few people nowadays make their own furniture or would freeze to death if they failed to gather firewood. Three generations ago local woodlands were still useful, four generations ago they were essential to life itself. It is no wonder the old forest rules were strictly enforced and, despite what we sometimes read to the contrary, those who earned their living from the greenwoods of Britain were more conservation-minded than we are today. Almost all early industries depended on wood. Some of these will be mentioned when I describe uses of the various trees later in the chapter, six of the old woodland industries (bow and arrow making, charcoal burning, chair bodging, besom making, hurdle making and tanning) are described below.

Bows and Arrows

It is difficult for us to appreciate how dramatically the invention of the bow and arrow must have altered the face of war. At first, archers simply waited at a distance and picked off the enemy spearmen at will. This was fine until the other side got wise and commissioned their own craftsmen to produce superior bows, strings and arrows. Then the would-be archer turned to Mr Bowman or perhaps Mr Bowyer (many such surnames owe their origin to country crafts) to fashion him a bow and to Mr Fletcher to make him arrows. Armed with his bow, he no longer felt like David fighting Goliath. He was now a mighty warrior.

In addition to their use in war, bows and arrows were also important in providing food for a hungry family. The traditional wood for the bow was yew, while the bowstrings were made from flax or hemp and the arrows from ash. But in England yew seldom grows straight enough to make good bows, so when England was not at war with France suitable lengths were imported. This supply was so vital that in the fourteenth century a curious tax was introduced. Every barrel of wine brought from France had to be accompanied by a number of stout yew timbers.

Charcoal Burning

Not far from my ash tree, the little witch and I were once picking blackberries when we found a bramble bush growing out of and half-concealing an old wall. We scraped around and uncovered an old potash pit where bracken had once been burned to produce alkali. We unearthed enough to see where the air used to enter and where the heaps of vegetation were loaded. A number of years later, I returned in the winter when the vegetation had died down. I once more found the potash pit and nearby was a fascinating-looking circle where, on digging down, I found charred pieces of wood. Subsequently, when I rang a friend I was at school with by the name of Ashburner and asked her what her grandfather, who lived in the neighbourhood, did for a living, she told me that he had made charcoal.

Charcoal is wood which has been roasted to drive out all the moisture, leaving behind only carbon. If the charcoal has been properly made it is hard and brittle, and when tapped has a ring like metal. These days charcoal has fewer uses, but it is still required by artists and is important in the manufacture of air filters. It has also been used in the produc-

Charcoal burning

tion of artificial silk, penicillin, fertilisers and deodorants and it is now sold in considerable quantities for garden barbecues. In the days of cannon and musket, the manufacture of gunpowder required a constant supply of charcoal. Also, as the industrial revolution gathered pace, ever increasing supplies were needed for smelting iron and great inroads were made into British woodlands.

Because of the difficulty and high cost of transport, charcoal tended to be produced on site and the burners lived in the woods for much of the year. They constructed a wigwam-shaped hut from timber and whole families worked the pitstead, which was the name given to their heap of burning timber.

The species of tree used for making charcoal varied from one part of the country to another. Alder-buckthorn and alder were used in some places, while in the Lake District oak, ash, beech and birch were used. The precise method of preparing charcoal also varied, but the basic principle was always the same. It was soon realised that timber stocks were not inexhaustible and from about 1500 the sensible practice of coppicing was in operation in the majority of charcoal-burning areas. Standard timber production can mean a harvest only every 60–100 years or so, but in coppicing the base of the trunk is left alone and some of the straight branches are cut every 13 or 15 years. The cutting of suitable timber began in the spring and it was stacked until the autumn, when burning began. The pitstead was prepared by levelling a circle about 20 metres in diameter and ensuring that it was properly drained. A stake was placed in the centre and the larger pieces of timber, usually about 1 metre high, stacked against it. The smaller branches were stacked and once the heart of the pitstead was built smaller, thinner pieces could be added. At this point the original stake – called the centre peg – was cut level or removed altogether, a thicker stake called the 'motty peg' hammered in and more sticks were leaned against this so that a smooth shape was produced.

Once all the wood was stacked, the skilled burners went round and looked for gaps through which air might enter. The pitstead was then covered with grass or reeds followed by 'sammel' – fine sifted subsoil – which ensured that no air could enter. All was now ready for the controlled burn. The motty peg was removed and burning charcoal dropped into the hole, which was then plugged with a sod. This ensured that the stack burned from the top down so that the steam and gases given off did not wet the unburnt wood. During the burn, which lasted three or four days, the pitstead tended to shrink and collapse. Then any breaks through which air could enter had to be sealed with damp grass and sammel. Breaks could occur at any time, day or night, and tending a pitstead was an exhausting business. Gradually the stack of wood

was reduced to embers and smouldering brands began to poke out from the bottom. These were chopped away with a sharp spade and the gaps sealed with sammel so the burn could proceed for a little longer. As the process neared its conclusion the hard-working burners could still lose the whole of their charcoal by one careless or clumsy action. Inside the stack it was so hot that if air was allowed in the charcoal would have burst into flames, ruining the whole enterprise. One burner therefore removed the sammel with a rake, whilst his mate poured in water. It was then sealed and another area opened and watered. This damping down took a further twenty-four hours and care had to be taken to use the correct amount of water, otherwise soft mushy charcoal was produced and the profits reduced accordingly. Experience certainly helped, but the old charcoal burners learned quickly – for a small error might mean that their families went hungry.

Charcoal burning was very demanding of timber, since seven tons of wood were needed to produce one ton of charcoal. However, the discovery of coke finally killed off the industry and probably saved what was left of our native woodlands.

Chair Bodging

I once knew a chap named Bill Bodger who was a forester. Although we have long since lost contact, I often wonder if he could trace his ancestry back to the bodgers of the beech woods, one of the most primitive of furniture crafts. Bodgers have been living and working in the Chilterns for centuries, producing legs, spars and stretchers for Windsor-type chairs. Beech has the advantage of being easily turned even when green and the bodgers, who work in pairs, do not have to wait for their timber to season. One member of the team cuts the beech

A bodger turning chair legs on a pole lathe

into suitable pieces, while his mate operates the odd-looking but highly efficient pole lathe, which is the oldest known rotating tool for turning wood. It consists of a treadle and a long springy larch pole, with one end anchored and the centre balanced over a cross-piece, leaving the other end free. This is linked to the treadle by a hemp rope, which is twined a couple of times round the wood to be turned. Pressing the treadle allows the bodger to work on the turning wood, the chisel being removed on the recovery stroke when the foot is lifted from the treadle. The speed and accuracy of a bodger using his deep gouge, curved, parting, skew and broad chisels is amazing and in the heyday of the craft a team could produce almost a thousand chair parts in a week. This was very demanding on timber – and when one area had been exhausted, the team moved off to a new area and set up their work tent and pole lathe there, allowing the original area to recover. Each team had the necessary incentive to conserve stocks because sons were expected to follow in their fathers' footsteps, just as they themselves had done. Each bodger also developed his own style of carving, so that his work could be recognised by the discerning eye.

Besom Making
It was not only the boys of the family who were fascinated by my great-grandmother's green cunning. She once spent a whole day in the woods making broomsticks for two girl cousins who had come for a holiday. I still find a home-made besom the best implement for sweeping up autumn leaves from the garden paths or a wet lawn. They are so simple to make that you might think no expertise is involved. The skill lies in making a besom to last, and the real expert chooses his timber with great care and no little skill.

Seven year old birch is cropped in the autumn in such a way that the tree will crop again after another seven years or so. The twigs are stacked in bundles and arranged crossways so that they look like a miniature hay stack. After over wintering the bundles are ready for use and are cut to shape, the unwanted twigs being sold as faggots for firewood.

It is fascinating to watch a broom-maker sitting astride his 'besom horse', which instead of a head has a vice for holding the twigs. This vice is operated by the 'rider's' foot while the twigs are being bound. In the Lake District the old besom makers preferred birch or hazel, but in some parts of England oak, ash or even willow was used. Latterly the twigs have been bound with wire, but in the old days the wild wood provided the necessary material in the form of blackberry stems shaved to remove the prickles, while the material for the handle depended on available timber. For my own part, I have always used ash, but lime and hazel have their devotees and the little witch, always an individualist, preferred blackthorn. Like many woodland industries besom making almost disappeared, but it is now making a return, mainly as a small cottage industry supplying local garden centres. Providing care is taken to coppice and also to replant the necessary timber, this must be regarded as a most welcome addition to the country scene.

Hurdle Making
Many of the early woodland industries involved selecting the right tree, felling it and then sawing up the timber on site. Large planks were produced by a sawyer and his assistant in a saw-pit about 5 metres long, with the timber arranged along it. One man took up position in the pit, whilst the gaffer was above ground. The wood was then cut up with a pit-

Horse chestnut – often used for making gates, hoops and yokes

saw. This method was employed in Britain, but in other parts of Europe trestles were used. As in most of the woodland-based industries, the work was done in summer. In the winter the sawyers travelled about the country, calling on wagon makers, coopers and wheelwrights and cutting timber shapes to order. Masons and carpenters also made frequent use of their services.

In a countryside heavily biased towards sheep, hurdle making was important so that stock could be 'folded' into suitable grazing areas. Two basic types of hurdle were produced – wattles, made from hazel rods, and gates, usually made from ash, which were often larger and more permanent structures. If hazel and ash were either not available or temporarily in short supply, then other species such as willow or horse chestnut were pressed into service. Hurdle making, rake making and the fashioning of walking sticks, hoops, yokes, shovels, forks and spoons could all make use either of coppiced timber or the unwanted strips produced by the sawyer as he cut up the standards in his saw-pit. Even the bark from fallen trees had a use and was eagerly sought by the tanners, who often had the surname Barker.

Tanning

At one time, especially in Wiltshire, 'bark harvest suppers' were held after the spring crop of bark had been gathered in the month of May, when the sap was rising. Oak was the tree favoured by those who practised 'ringing', which entailed skilfully stripping bark from a tree, without killing it, so that after a period of recovery the tree could be cropped again. As a result, early tanneries were usually situated either in or close to oak woods. The Forest of Dean boasted a number of notable tanneries, but there were many small concerns operating throughout the country, especially in the Lake District.

First of all, the oak bark and even acorns, which are also rich in tannic acid, or lime bark (also popular with tanners) were ground to a fine powder in a bark mill. The powder was then added to water in a vat called a leaching pit in which the hide was immersed. There were a series of leaching pits, each stronger in tannic acid than its neighbour, and by the time the hide was removed from the last it had been converted into leather. Given the multitude of uses of leather before synthetic substitutes were invented, tanning was one of the most important of all the woodland industries.

These craftsmen were not alone in their dependence on wood. All country folk visited woods to gather firewood and edible or medicinal plants. The old forest laws did not permit the common people to cut down trees for firewood, but they were allowed to collect any timber they could reach with a 'hook or crook'. Hence the origin of 'by hook or by crook'. The little witch taught me to look at every tree, flower and plant with an eye to what it could be used for. Some of these uses are described in the remainder of this chapter.

Forest Trees

As John Evelyn pointed out, a nation would be better off without gold than to be short of trees. Nowadays oak, beech, ash and birch are the most important forest trees in Europe and hornbeam has been an important crop tree in some areas.

Oak *Quercus* spp

As we have seen, the oak has played a major part in forest industries and

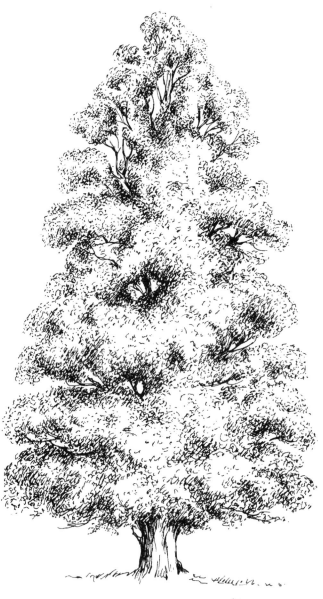

Lime – the bark was often used for tanning

country crafts. There is still no better material for ladder rungs and wheel spokes than cleft oak, which has tremendous strength since the grain refuses to be broken.

As a teenager walking in the Lake District, I once stopped to talk to an ancient dalesman who seemed to be boiling oak logs outside his cottage outhouse and soon learned that the wood thus softened could be easily split into thin bands, then woven into tough, long-lasting baskets called spelks. He spoke of his grandfather, who had hailed from Kent and had spent most of his life making cleft-oak pegs, which were then much used in joinery and for securing roof tiles to laths. In the days when timber was cheap the laths themselves were of oak, as were the huge rafters, and in the southeast of England the shingles used to cover church steeples were made of oak too.

Making a spelk basket

Coopers demanded cleft-oak staves for making beer, wine and whisky barrels. When whisky is first distilled, it contains a very high alcohol content and is almost colourless. As it matures, alcohol evaporates through the pores of the oak cask and the typical pale colour is absorbed from the barrel. Indeed, whisky is often matured in old sherry barrels in order to improve the colour. As the alcohol level drops and the colour deepens, the mellow taste of the whisky improves too. I may be wrong, but I suspect that the flavour of real ale may well come as much from the barrel as from the brewing.

Great demands were made on English oak during the late sixteenth century after Henry VIII's decision that England needed a navy – a policy which bore fruit when his daughter Elizabeth I was able to defeat the Spanish Armada with 'stout ships of oak'. As England became a prosperous nation, wooden merchant ships and half-timbered mansions and other buildings were constructed all along the coast and throughout every county.

Two species of oak are native to Britain, the durmast or sessile oak (*Quercus petraea*) and the common or pedunculate oak (*Quercus robur*). The acorns of the sessile oak have no stalks and seem to sit directly on the twig (the word 'sessile' comes from the Latin meaning 'seated'). In contrast, the leaves have quite long stalks. In the common oak the situation is the reverse – the leaves have no stalks, while the stalks of the acorns are quite long. Generally, the former grows straighter and provides the better timber. The two species are certainly inter-fertile and oak woods often contain trees that are a cross between the two.

The word oak derives from the Saxon 'ac', hence the word 'acorn'. At one time acorns were ground to make flour, but by the sixteenth century the standard of living had obviously risen and Turner was

Common or pedunculate oak

able to write, 'Oke, whose fruit we call an acorn or an eykorn (that is ye corne or fruit of an Eike) are hard of digestion and norishe very much, but they make raw humores. Wherefore we forbid the use of them for meates.'

The oak was venerated by the Celts, who called their holy tree *derw* and their priests Druids, while the ancient Greeks called the oak *dryas* and wood nymphs dryads.

The Oak and the Druids

More nonsense has probably been written about the Druids than about any other religion. There is no doubt, however, that theirs was a 'green religion', with a heavy emphasis on the wildlife of the forest. As the most majestic tree in the forest, the oak was inevitably the chief object of veneration. In their book *Chapters on Trees*, published in 1873, Mary and Elizabeth Kirby described the part played by the oak in the 'old religion':

It was an article in the Druid's creed, and one to which he strictly adhered, that no temple with a covered roof was to be built in honour of the gods. All the places appointed for public worship were in the open air, and generally on some eminence from which the moon and stars might be observed, for to the heavenly bodies much adoration was offered. But to afford shelter from wind or rain, and also to ensure privacy and shut out all external objects, the place fixed upon, either for teaching their disciples, or for carrying out the rites of their idolatrous worship, was in the recess of some grove or wood. These groves were planted for the purpose, and were always composed of the tree most precious to the Druid, the old British Oak. So high was the

regard shown to it that the Druid priest assured the people the Oak was the favourite of the gods, and everything belonging to it was sacred. And he would undertake no ceremony or rite without the Oak chaplet being bound on his forehead: The sacred grove was usually watered by a river or stream, which was also considered holy. And here, in some open space, with the great branching trees spreading around, and large stones set upright; and sometimes one huge stone would lie in a horizontal position on two of its fellows, and form a stone table, which was the primitive altar on which the Druid priest offered sacrifice. This stone table was called a 'cromlech', and upon it at a certain time every year was kindled the sacred fire of the Druids, and here burned the famous yule log, the name of which is familiar to us even at the present day. The sacred fire was said never to be extinguished, but to continue burning, under the care of the Druids, from one generation to another.

But on the last day of October, when winter was close at hand, one of the most important ceremonies connected with the pagan worship of our ancestors took place all over Britain. The people were compelled to extinguish their fires, and present themselves in the sacred groves with a certain payment in their hands, the perquisite of the Druids. This act of worship was insisted upon with the utmost rigour, and if from any cause it was neglected, the most heavy penalty was inflicted. The rebellious person was laid under a kind of interdict, and shut out from all intercourse with his fellows, and not allowed to kindle a fire in his dwelling. When, however, the obedient populace had done according to the law, and paid the due offerings to the priests, they were allowed to kindle a log from the sacred yule log of the Druids, which was burning on the altar.

This yule log was always of Oak, and for centuries afterwards, its memory and the superstition connected with it lingered in our midst. It was the custom in many households to throw a log on the fire at Christmas, and take it off again, to be reserved until another year. Thus the old custom long survived the fall of the Druids. The great yule feast of the nation was held in honour of Baal or Bel, who was god of the sun and of fire, and from whose name the word was derived. It took place in midwinter and not only do we still have the Yule log as part of our Christmas decorations but the Mistletoe, another of the Druids' plants, is also used.

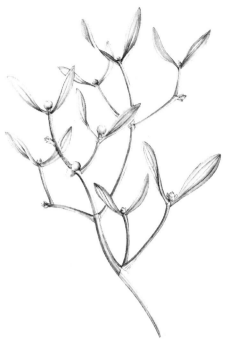

Mistletoe

Mistletoe *Viscum album*

This fascinating plant is a partial parasite which produces suckers that grow into the tissues of the host tree. Although mistletoe is green and makes some of its food by photosynthesis, a large proportion of its essential carbohydrate is extracted from the host parasitically.

The mistletoe's method of growth is simple. The branches subdivide continually, each new fork representing a year's growth. Each of the two stems that make up a new fork terminates in a flower bud, flanked by a pair of leaves, and two side buds. From these the stems develop which make up the next fork, and so on. A cluster of flowers sits snugly in the 'V' of each fork, male and female flowers being borne on separate plants. The white berries are produced after the female plant has been pollinated and each contains only one seed, surrounded by a sticky substance which deters many birds from eating them. A notable exception is the mistle thrush. After a meal the thrush wipes off the excess 'glue' by rubbing its bill on the bough of a tree, thus providing a second method by which the plant is spread.

The Druids were not alone in thinking that mistletoe had magical powers. Indeed, the family to which our own mistletoe belongs was venerated in Africa, Japan and New Guinea. As the arrangement of the berries resembles the male sexual apparatus, they played a vital part in fertility ceremonies both in England and in other parts of the world, and women who wished to conceive wore mistletoe round their necks. An echo of these ceremonies still survives when we kiss under the mistletoe at Christmas.

The berries were at one time eaten in an attempt to cure epileptic fits. This is not quite so outrageous as it sounds, because the whole of the plant is permeated with an anti-spasmodic substance that also reduces blood pressure. Like many of the ancient remedies, this treatment may well have killed as many as it cured, since mistletoe contains the chemicals tyramine and betaphenylethylamine, which in quantity cause gastroenteritis and heart failure, though they do not seem to affect the mistle thrush.

Mistletoe thrives on apple trees, and in Somerset and other parts of England it is cultivated in orchards for the Christmas market. It has also been found growing on hawthorn, maple, poplar and lime, but rarely on oak. The Druids' ceremonies demanded that the mistletoe used in the rituals grew on oak. Mary and Elizabeth Kirby therefore suggest that in order to ensure that there was never an embarrassing hitch, the priests may have been guilty of sleight of hand:

Another of the Druid ceremonies was to go in search of the Mistletoe, growing, as it very rarely does, upon the Oak. For the Mistletoe was as highly venerated as the Oak itself, and was much used in their worship. This feast was held in the early spring, which was the new year by the Druids' reckoning of time. And sacrifices of human victims which make the remembrance of them and their religion to be abhorred. Huge baskets of Oak twigs woven together were made to enclose the wretched beings who were to suffer and who were generally criminals, or prisoners taken in battle.

When all this was done, the Druid priests, clad in white and flowing garments, issued forth to search for the Mistletoe amid the trees of the sacred grove. No doubt the search had already been privately made, and the spot indicated to them. But they always pretended that a vision had revealed it. At any rate, when the procession arrived at the Oak on which the parasite was growing, much pomp and ceremony took place. Two white bulls were bound to the tree by their horns, and a priest ascended to the branch where the Mistletoe flourished, and severed it with a golden knife, while another priest stood below and held out his white robe to receive it. If it fell to the ground a great calamity was thought to be impending.

Next to the Oak the Apple-tree was held in reverence, and orchards were usually planted near the Druids' grove. This was a favourable circumstance for the propagation of the Mistletoe, since it grows abundantly on the Apple-tree, and the seeds were likely to be wafted by the winds or carried by the birds to the neighbouring Oaks.

In those Druidic days criminals were tried in the open air. The judge was seated under the Oak, with his primitive jury beside him, and the culprit stood in the midst of a circle made by the wand of the Druid. We might also mention that the well-known chorus, so often heard amongst us, of 'Hey derry down,' is said to be nothing more than a corruption of an old Druidic chant, and literally means – 'In a circle the oak moves round'.

Other religions used mistletoe with a little more gentility. Followers of Friga, the Scandinavian goddess of love, kissed under the mistletoe and several European cultures used it to ward off witchcraft. Even Culpepper, following the writings of Clusius, noted that 'Mistletoes, being hung about the necke reduced Witchcraft, but it must not be allowed to touch the ground after it is gathered'.

What secrets the ancestors of our venerable oaks could tell if only they could speak! Whilst the folklore surrounding the oak is intriguingly difficult to unravel, some aspects of its natural history are equally complex. Not least amongst these wonders are galls.

Oak Galls

At one time ink was produced commercially by crushing oak galls (particularly oak apples), which like acorns are rich in tannin, together with salts of iron. I have no doubt that the idea developed from the observation that when an iron tool is leaned against a freshly cut tree the action of the tannin on the metal produces a black stain.

Originally the making of books was wholly based on woodland materials. To make a book, leaves were boiled to remove the colour and were bound between two pieces of bark or wood. Sometimes the leaves were stitched in, or stuck in with glue made from bluebells. All that was now needed in order to produce a serviceable book was to replace the animal blood often used in early days with ink that would not fade. Ink made from oak galls or from the ink-cap fungus provided the perfect answer.

Oak apples and other galls have nothing to do with the reproduction of the tree. When a plant is attacked by a parasite it may react by enveloping the offending organism with a layer of tissue, effectively sealing it off from the rest of the plant. Some interesting research is being carried out at the moment to identify the chemical causing gall formation. However, it is even more important to discover the nature of the chemical which stops the growth, since it may well have a use in the treatment of cancer.

Many galls result from the activity of chalcid wasps, which have a fascinating life history. They usually overwinter in the soil under the tree, and in spring the wingless females emerge and crawl up the trunk. They are fertilised by the winged males and the eggs are laid in the young buds. The gall is then formed around the egg and provides food for the developing grub. When ready to pupate, the grub bores a hole in the gall and makes its way down the trunk and into the soil. The oak apple, or other gall, then rots away and its rich tannin content can be used to cure leather or make ink.

Beech *Fagus sylvatica*

Easily recognised by its long slender pointed buds, the beech has always been popular with carpenters because the timber does not warp, a fact well known to the bodgers. It also seems that the word 'book' may be derived from the Old English word meaning beech, since early scribes often carved their mystic runes upon blocks of beech. The little witch collected beech twigs just before Christmas and, sitting quietly after a meal, would use the pointed buds as tooth picks. Many years later, when I read the works of the seventeenth-century diarist John Evelyn, I found he recommended beech buds as toothpicks, provided they were 'winter hardened and dried upon the twigs'.

The little witch loved watching the chaffinches and bramblings feast on the beech nuts before taking her share of the crop. Beech does not grow all that well in Britain and some botanists even doubt whether it is native. I believe that it is native, but requires chalk-rich soil to thrive and only finds ideal conditions in the south-east of Britain. The timber is so good, however, that beeches have been planted throughout the country and they can sometimes be grown in less than ideal conditions if Scots pines are used to shelter the young trees. It is said that Queen Victoria insisted on beech logs

Beech

being used on the fires at Windsor, as they produced such a hot flame.

Because beeches are often grown in an unsuitable climate, the triangular nuts do not always mature within their rough, spiny cases. When our own beeches produced a crop the little witch knew it was time to polish the furniture. She gathered the nuts, wrapped them in a duster and pressed them with a rolling pin. Soon the duster oozed with oil and this she rubbed on our furniture, which – much to my mother's alarm – immediately turned green. After a vigorous polish, however, the green colour was replaced with a water-repellent shine. My mother told me that one of her mother's relatives had been in service at a stately home and they used nothing but beech oil for polishing. We still use it on our furniture to this day.

Beech timber does not last well out of doors, but it can be sliced into thin, broad bands. These are subjected to steam, which softens the wood so that it can be bent into shapes suitable for making barrels and the rims of sieves, boxes and drums. Damp air rots beech wood, but it lasts well when permanently waterlogged. The supports of the old Waterloo Bridge in London were of beech, as are the foundations of the magnificent cathedral at Winchester.

Few plants grow under beeches – partly because the leaves decay slowly, but also because, although the leaves are small, they overlap and the floor of a summer beech wood does not receive much light. This may well be one reason why beech has proved popular as a hedge tree, since the area beneath needs very little maintenance. Another advantage is that it is not affected by clipping and almost certainly the finest beech hedge in the world is at Meikleour on the Perth to Blairgowrie road in Scotland. It soars to a height of 30 metres and runs for a quarter of a mile, with over 600 trees standing straight like soldiers on parade.

In the Middle Ages it is doubtful if beech was used for anything except firewood and the excellent charcoal which can be made from it. The heyday of the beech had to wait until there was a demand for household furniture and for sophisticated metal tools with which to fashion it. Beech woods were also the source of truffles. Truffles were still collected commercially in Dorset until 1925, but the skill required to find these delicious fungi underneath the rotting beech leaves has now vanished in Britain, although truffle hunting still goes on in France, Germany and eastern Europe. Pigs can smell out truffles, but naturally enough gobble them up themselves unless they are muzzled. Trained terriers are therefore used nowadays and when rewarded with juicy meat make more efficient truffle hunters.

Ash *Fraxinus excelsior*

I used to wonder why the little witch planted an ash rather than an oak for me, but few trees have such resilience and strength as this sacred tree of the ancient Scandinavians. In the Scottish Highlands the influence of the Norsemen was evident until fairly recently. Newborn babies had ash sap squeezed into their mouths in the hope that they would acquire the strength of the tree. The Norsemen believed that the ash, which they called *yggdrasil*, was the strength of the world, with its roots in hell and its branches in their equivalent of heaven. The trunk and lower branches were in our world and messages were carried to and fro by the agile squirrel. In the early days of Christianity the priests, at pains to stamp out the old religions, organised red squirrel hunts to prove that the animal was mortal and was not under the protection of false

Ash

gods.

The idea that the ash was a healing tree survived even longer and died out almost within living memory. This involved the village shrew ash. Shrews were believed to cause lameness in both cattle and people who crossed their path. When a child was born with a deformed or defec-tive leg, a shrew was blamed and the only cure was to kill the shrew and beg strength from the ash tree. With great ceremony a shrew was captured and a hole was made in the trunk. Then the shrew was dropped into the hole and the wound in the tree was bound up. It was believed that as the wound healed the child would gain strength. In his *Natural History of Selbourne*, published in 1789, Gilbert White describes a shrew ash which could cure animals with injured limbs:

> There stood about twenty years ago a very old, grotesque, hollow, pollard ash, which, for ages, had been looked on with no small veneration as a shrew ash. Now a shrew ash is an ash whose twigs and branches, when gently applied to the limbs of cattle, will immediately relieve the pains which a beast sufferes from the running of a shrew mouse over the part affected; for it is supposed that the shrew mouse is of so baneful and deleterious a nature, that whenever it creeps over a beast, whether it be horse, cow or sheep, the suffering animal is afflicted with cruel anguish, and threatened with the loss of the use of its limbs. Against this accident, to which they were continually liable, our provident forefathers always kept a shrew ash at hand, which, when once medicated, would maintain its virtue for ever. . . . A shrew ash was made thus; into the body of the tree a deep hole was bored with an auger, and a poor devoted shrew mouse was thrust in alive, and plugged in, no doubt with several quaint incantations long forgotten.

Although the incantations have long been forgotten, we still make good use of the ash's strength. We may no longer kill our enemies with spears fashioned from ash boughs, but cricket bat handles and hockey sticks are still made from ash, as are police truncheons and the shafts of carts. Another present-day use is in the manufacture of oars and snooker cues, the smoothness of the wood being an additional advantage, and there are still a few craftsmen up and down the country who maintain pollarded ash trees from which they make walking sticks. I still have an ash walking stick given to me by the little witch. She told me that her father had made it, and I keep it handsomely polished with beech-nut oil.

Ash can be recognised in winter by its greyish bark, which in young trees is very smooth, and by its black buds (a feature shared by no other British tree). The twigs are flattened, especially close to the large terminal bud, and the other buds are arranged in opposed pairs. The leaves are equally distinctive. They are compound in structure, the central stem of each leaf carrying five pairs of leaflets with serrated edges and a single terminal leaflet. The ash, like the oak, is late coming into leaf and an old country rhyme, which does not seem to run true, states that:

Oak before ash, there will only be a splash.
Ash before oak there will be a soak.

The dark flowers, pressed close to the end of branches, appear in April long before the leaves. When ripe the purple stamens open like a spray and are dusted with pale yellow pollen. Each flower is bisexual, but the male part dies off before the green female part develops, thus ensuring cross-pollination. Large amounts of light pollen are blown about in the wind and a high percentage of female flowers are fertilised. The occasional ash tree produces male flowers only and these of course never bear fruit. By June the seeds, technically called samarae but more usually known as keys, are hanging in green bunches from the branches. Each key has a flat membrane twisted rather like an aeroplane propeller, with the seed at the end closest to the stalk. In autumn the keys twist and turn in the air and may be carried miles from the parent tree. Ash trees do not produce a good crop of fruit each year, but the little witch never failed to find an ample supply of green succulent keys, which she pickled in vinegar to make a tasty addition to salads.

Birch *Betula* spp

In 1842 J. C. Loudon wrote of the birch:

> The Highlanders of Scotland make everything of it; they build their houses, make their beds, chairs, tables, dishes and spoons; construct their mills; make their carts, ploughs, harrows, gates and fences and even manufacture ropes of it. The branches are employed as fuel in the distillation of whisky; the spray is used for smoking hams and herrings, for which last purpose it is preferred to any other kind of wood. The bark is used for tanning leather, and sometimes, when dried and twisted into a rope, instead of candles.

There is no doubt that the timber was essential to the lives of the Highlanders and Islanders. It should not be assumed from this that birch is the perfect timber – indeed, it is not easy to work – but on many mountainsides and almost all the Western Isles birch is the only species which grows in any quantity, though it does not live long and is subject to attack by fungi, especially bracket fungus and the colourful but dangerous fly agaric.

Bracket Fungus *Polyporus betulinus*

Bracket fungus has thread-like hyphae (slender, branched filaments) that penetrate the tissues of the birch. The saddle-shaped fruiting bodies, often over half a metre across, are in many parts of the country called dryad's saddle and illustrations in books of fairy stories often depict fairies sitting on them. Another name is razor-strop fungus. In the spring and throughout the summer the saddles are soft and rubbery, but in autumn they become harder. At this time they used to be harvested and cut into strips for sharpening razors and the tools used by those who earned their living from the woods. These have now gone the way of the cut throat razor, but another use still has a few adherents even today. Bracket fungus does not decay and is cut into blocks on which reference collections of insects can be mounted. The little witch had her own use for 'fairy fungus', as she called it. Small specimens she employed as pincushions, slightly larger ones she kept for darning needles.

Fly Agaric *Amanita muscaria*

In the days when hygiene was non-existent, the fly population around dwellings must have been horrifying and a fly-swat made of birch twigs could have brought only temporary relief. Another fairy fungus thriving in birch woods provided a much more effective remedy. The 'pixie's seat mushroom', known today as fly agaric, is easily recognised by its scarlet cap, often flecked with white. It looks and is extremely poisonous, but always seems to be surrounded by flies. After giving us all a careful warning, the little witch used to collect fly agaric and stand the fungus in old saucers which she kept topped up with water. These were placed on the window-ledge of the kitchen and on the bench by the side of the old privy at the end of the garden. This was of course only possible in late summer and early autumn, when the fungus was available, but this was the very time when the fly populations were at their highest, especially after a hot summer. In the autumn of 1947 I remember seeing stupefied flies falling from the agaric and drowning in the water in the saucers – a

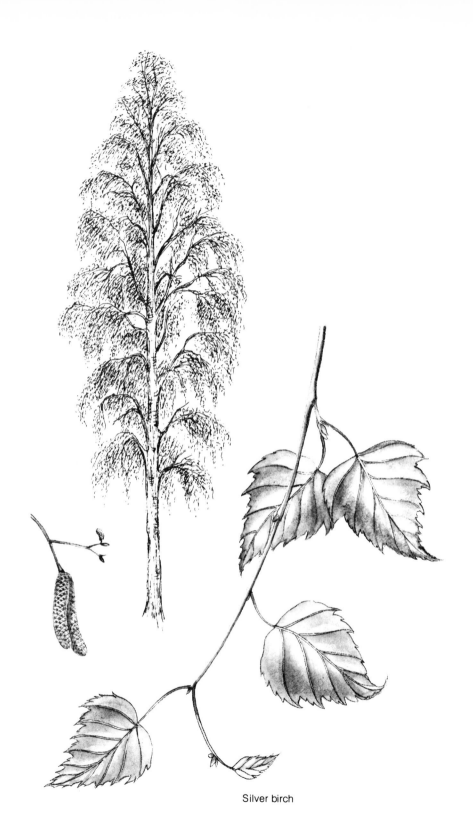

Silver birch

method of disposal documented as early as the thirteenth century, when containers of milk with chopped fly agaric in them were placed in all the rooms of important houses. The practice was common throughout Europe and in some of the poorer areas it continues to this day.

The Uses of Birch

In our woodlands we have two species of birch, the silver birch (*Betula pendula*) and the common or downy birch (*Betula pubescens*). In Britain silver birch, with its peeling slivers of bark which made good kindling for camp fires, tends to dominate southern woodlands. The tapered, serrated leaves are pale silvery green on the upper surface and bright green below. Their delightful scent fills the whole wood when they are fanned by a gentle breeze after rain.

The common birch occurs more widely than the silver birch in northern areas. The bark is seldom white and it has downy, purplish-brown twigs. When the sap starts to rise in spring, the twigs become greasy and the liquid produced by boiling them in water can be used as a hair rinse or shampoo. The oily substance that bubbles out when the bark and twigs are warmed over a gentle fire was once used for polishing leather and I have found it to be most effective both for shoes and for leather-bound books. Because of its high tannin content, in times gone by birch bark was used for wrapping perishable materials, especially dried food, prior to storage. Several birch bark containers dating from about 5,000 BC were unearthed at Starr Carr, near Scarborough, and it is thought that Neolithic hunters made a powerful glue from the bark for securing the stone spearheads to the shafts of their spears. My great-grandmother used to obtain this sticky fluid from a healthy birch in spring by

Fly agaric

boring a hole and patiently collecting the running sap in a cup to serve as the basis of her very potent birch wine.

Most of the early uses of birch, including the provision of the oil burned in lamps, have long since disappeared, but the timber is still used in the production of plywood. Thin layers of birch are glued over a central layer of spruce with the grain laid at right angles, which gives it great strength. With care small pieces of birch can be turned and smooth birch-wood cotton reels were once produced in huge quantities in bobbin mills up and down the country, especially in the Lake District. Plastic has now taken over, as a cheaper alternative, but at Scott Park, near Newby Bridge, close to Lake Windermere, an old mill has been beautifully restored and the charm of this birch-dependent industry can be appreciated.

The birch also played a part in the old tradition of 'brush marriages'. In the wild north of England, when the great forests

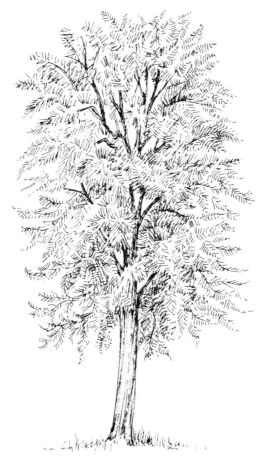

Hornbeam

still existed, couples wishing to marry often had to wait a long time for the blessing of the Church. One solution was for the couple to stand together in the presence of their families in front of a birch besom. They then held hands and jumped over the brush. Once they had jumped the broomstick it was regarded as permissible for them to live together 'ower t'brush' until a priest made the marriage official.

Hornbeam *Carpinus betulus*

In his seventeenth-century herbal John Gerard, who was born in Cheshire, wrote that hornbeam 'growes great and very like unto the elme or wich hasell; having a great body, the wood or timber thereof is better for arrowes and shafts, pulleys for mills, and such like devices, than elme or wich hasell; for in time it waxeth so hard that the toughness and hardness of it may be rather compared unto horn than unto wood; and therefore it was called hornbeam or hardbeam'.

Considering that it is one of our native trees, surprisingly little is known about the hornbeam. Some parts of the tree resemble birch, others beech or even elm. As a result, it is not always easy to recognise. Hornbeam was common in the forests of southern Britain, but unless planted specially it is something of a rarity in the North.

As the above extract from Gerard indicates, the value of the tree lies in the hardness of its timber. The ancient Greeks knew it as *zugia*, which was also the word for 'yoke', and in many parts of Britain hornbeam was traditionally used for making yokes for ploughing and carrying milk churns until well into the nineteenth century. The wheelwright appreciated its value too, as did the makers of wooden gear cogs for mill machinery.

Hornbeam can reach a height of more than 20 metres. The fluted, elliptical trunk fans out into a flurry of often overlapping branches and the smooth, dark grey bark is somewhat reminiscent of beech. The long, pointed red-brown leaf buds curve inwards towards the twig – a characteristic unique to hornbeam – and the flower buds open in April. The pendulous male catkins are greenish in colour and are dusted with yellow pollen when ripe. The female catkins are carried on the same tree but usually grow nearer to the end of the twig and ripen after the males, thus preventing self-pollination. When fertilisation has occurred, flattened green seed-capsules, popularly called 'lanterns', develop. These are obvious from August onwards and are fully ripe by October.

The oval leaves, about 6 cm long and 2½ cm broad, ending in a sharp point, appear during April. The margins are toothed and the whole leaf has prominent veins. After the leaves have changed to their coppery-brown autumn colour they often remain on the tree until the following spring.

6 Woodland Shrubs and Herbs

A botanist's view of a deciduous wood reveals four distinct layers. First there is the canopy produced by the leaves of the dominant trees. Below this, in descending order, is the shrub layer, consisting of smaller trees and shrubs; the field layer, made up of flowering and non-flowering plants; and the ground layer of liverworts and mosses.

The Shrub Layer

I remember one particularly enjoyable ramble through a wood near our home with my great-grandmother, intent on gathering hazel nuts and rods. The rain had just stopped and water dripped from the autumn leaves of the high oaks and ashes on to the understorey of hazel and elder, both heavy with fruit. A nearby holly was just showing signs of forming berries in preparation for Christmas and a particularly strong-looking ivy plant had a stranglehold on a dying birch. Hazel, holly, ivy and elder all had their folklore and uses, some of which are still relevant today.

Hazel *Corylus avellana*
Apart from producing delicious and nutritious nuts, the hazel has been an important source of flexible sticks for making fences. At first the wild woods offered liberal supplies, but gradually, as keeping domesticated animals became more common, the supply of hazel rods was threatened by browsing cattle. As early as 1483 in the reign of Edward IV, steps were taken to safeguard supplies and

an Act was passed to authorise the enclosure of hazel woodlands 'with sufficient hedges to keep out all manner of Beasts and Cattel out of the same ground for the protection of their young spring'. Even then the hazel was coppiced every seven years and the practice must already have been long established. Indeed, a glance at a map will show where some of these old coppices were situated. The Saxon word *haesil* clearly accounts for the name Haslemere in Surrey. The Welsh word for hazel was *collen*, from which we get Llangollen, and Cowdray Park in Sussex derives its name from *coudrier*, the Norman word for the tree. Many of these early coppices have survived because when they became seriously threatened in the middle of the nineteenth century gamekeepers recognised their importance as cover for pheasants and partridges, whilst huntsmen knew that many a fox's

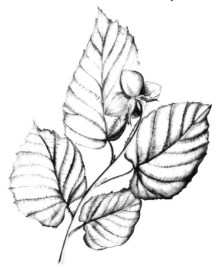

Hazel

earth is shaded by a hazel. Another factor which helped save the hazel was that it is the only British tree to produce regular crops of edible nuts. In parts of Kent hazel orchards produce two types of nut – cobs and filberts. The latter have long bracts hanging from them and may once have been called 'fullbeards'. Leicester City football ground is called Filbert Street and almost certainly was once part of an ancient woodland with an under-storey of hazel.

I spent many a happy hour with my great-grandmother gathering hazel rods, which she divided into 'dead 'uns' and 'live 'uns'. The dry dead twigs were pushed into the oven and were soon crackling and baking the bread. To the top of the dough she often added a sprink-ling of crushed hazel nuts, sometimes supplemented with a crushed acorn for good measure. The more supple twigs she wove into quite serviceable shopping baskets, the gaps between the sticks being filled with twists of birch bark. She used the technique once employed for making hurdles for fences and for the walls of houses, which were plastered with mud, cow dung, hay or horsehair, depending on what was available when the wall was being built.

She once took me on a long walk up a steep hillside to see a local farmer whose coughing cow had been cured by her cowslip and comfrey lotion. He was just on his way out into one of his fields with the local dowser, or water diviner, to look for the best place to dig a small drinking pond for his stock. The small, thin, wrinkled man closed his eyes and held a forked rod of hazel across the palms of his hand. He walked slowly around the field, then suddenly twitched as the dowsing rod lifted apparently of its own volition. 'There,' he grunted and held his left hand out for the farmer's ten shilling note, his right pointing with the stick to a patch of dry ground. Two hours later a hole had been dug and was already filling with water. On the way home we cut several forks of hazel, none of which worked. But my disappointment was soon overcome when I found a straight rod over 2 metres long from which we made a fishing rod still in use today. Dowsing with hazel rods was thought at one time to be able to discover other things besides water, including valuable minerals, buried treasure and criminals guilty of murder. John Evelyn in the middle of the seventeenth century wrote that 'dowsing when examined by learned and credible persons, who have critically examined matters of fact, is certainly next to a miracle, and requires strong faith.' Indeed, it still does!

The bark of hazel is smooth and brown. The winter twigs have scaly olive-green buds arranged spirally along them, and the 'lambs' tails' which swing in the wind and discharge clouds of pollen are among the first welcome signs of spring. The tiny crimson female flowers are carried on the same shrub, but they do not ripen at the same time as the male cat-kins and you need to take a close look to see them. Gradually the fertilised female flower develops into a nut, which is usually ripe and ready to eat in October but sometimes as early as September. In the old days nuts were in great demand on All Hallows Eve, which falls on 31 October and used to be known as 'Nutcrack Night'. The little witch knew the ceremony well enough and the girls in our family enjoyed collecting their hazel nuts, writing the names of the local boys on the shells, then placing them on the hob. The first one to pop bore the name of your future husband. I 'popped' for a plump girl named Ada Phillips – and haven't seen her since.

Toothwort *Lathraea squamaria*

Some trees have particular flowers associated with them and it is surprising how often toothwort is found growing on hazel.

If you look at toothwort, the first thing you notice is that it lacks leaves and no part of the flower is green. Consequently, it cannot manufacture its own food from sunlight and lives as a parasite, most often attached to poplar or hazel, although I have found it to be increasingly common on sycamore. The roots of toothwort form suckers able to penetrate the root tissues of the host tree and absorb nourishment from it. Recent research has confirmed that this is the toothwort's main method of feeding, though it may also be able to attract flies, which crawl inside the flower, become trapped and are digested by enzymes produced by the plant, so it seems likely that toothwort is semi-insectivorous during the time that the flower is above ground. It is equally versatile in its methods of reproduction. Normally the pollen is carried from the male anthers to the female stigma by bees, but if they fail to perform this function the pollen grains are light enough to be distributed by the wind.

The shoots of toothwort are covered with thick, pink, fleshy scales. Indeed, its botanical name is derived from the Latin *squama*, meaning a scale. The whole scaly structure bears a resemblance to a human jawbone complete with teeth and this led to its use as a cure for toothache, another example of the doctrine of signatures.

Holly *Ilex aquifolium*

> The holly and the ivy
> When they are both full grown
> Of all the trees that are in the wood
> The holly bears the crown.

This traditional Christmas song underlines the ancient association between these two evergreen trees. Henry VIII may have deserved his reputation as a bombastic tyrant, but there were times when the liveliness of his intellect shone through his darker nature. His love of the countryside is echoed in a song he wrote which included the lines:

> Ah! the holly groweth green
> With ivy all alone
> When flowers cannot be seen
> And greenwood trees are gone.
> Green groweth the holly and so
>   doth the ivy,
> When winter blasts blow never so
>   high,
> Green groweth the holly.

In Cornwall holly is called 'Aunt Mary's tree', in Suffolk 'the Christmas tree' and in Yorkshire 'Christ's Throne'. In Somerset it is called 'the crocodile', probably referring to the fact that it is used in the construction of long, low hedges.

Both holly and ivy were plants associated with magic powers and were deemed especially suitable for use as protection during the dead of the year, although they were by no means excluded from festivals at other seasons. The red berries of the holly were considered to be particularly potent since the colour itself was supposed to give protection against evil. Thus the rowan tree, with its scarlet berries and the red breast of the robin were also considered lucky.

The tough, prickly leaves of the holly are replaced gradually throughout the year, so the tree does not have an autumn 'fall'. Although the berries are conspicuous, the white blossom which occurs in May and June is often quite insignificant. Holly flowers are either male or female

Holly

and are often borne on separate trees, the male trees bearing blossom, but no berries. The pollen is carried from the male flowers to the female flowers by insects, usually bees.

Holly trees grow slowly and are long-lived, but they do not grow very tall and one of 15 metres can be regarded as a fine specimen. The timber was favoured by cabinet-makers, particularly for inlaying. The holly family is widespread throughout the world but only one species is native to Britain, where it is particularly common on the western side of the country, especially in the Lake District. Prior to the twentieth century bird lime made from the berries was exported to the East Indies, where it was used to trap not birds but destructive insects. Holly seeds have a long dormancy period and do not germinate until their second year. Although birds can eat the berries without ill effect, they are injurious to human beings (a fact which should be stressed to children), causing violent vomiting without actually proving fatal. Much of our wild woodland has been destroyed in the last two thousand years, a process which accelerated during the nineteenth century, as a result of the industrial revolution, and the earlier part of this century. If you wish to find out how common holly used to be in your city, town or village, look at an Ordnance Survey map. The old name for holly was 'holm' and the appearance of 'holm' in a place-name often indicates that the species occurred there in the past.

The holly is indeed a 'holy' tree and it is thought that the Romans sent gifts to their friends decorated with a sprig of holly to celebrate the festival of Saturnalia. This festival was celebrated only about a week before our Christmas, so the pagan ritual was absorbed into our celebrations. Pliny the Elder, the most competent naturalist of the period, called holly *acifolium*, which simply means needle-leafed, a description now incorporated into our scientific name for the tree.

Ivy *Hedera helix*

Ivy, too, has a variety of vernacular names used in different parts of the country. These include 'bentwood' and 'bindwood', both of which make good sense. 'Love stone', used in Leicestershire, is rather more obscure, though it may be that the seeds were used in ancient times as an aphrodisiac.

Ivy climbs by means of 'adventitious roots', which erupt from the stem and grow away from the light. The roots therefore grow into any available dark crevice and there hold the plant in position. Some assistance is given by the production of a cement-like substance which makes ivy difficult to dislodge. The leaves, like holly, are evergreen and their shape alters as they mature. Young leaves are three or five lobed, adult leaves are undivided.

The little witch always gathered ivy leaves before a funeral. She boiled the leaves in water and, when it had cooled, soaked her black silk dress in it. It came out looking like new.

The flowers are yellowish-green and are produced from adult branches in October, providing insects with a source of nectar when many other flowers have succumbed to the first frosts. Ivy attracts red admirals and often that other migrant butterfly, the painted lady. The flowers eventually produce oval black berries round about Christmas. On no account should the berries be eaten, since they contain a toxic chemical called hederagenia. Even a couple swallowed by a child can be dangerous and the consumption of several berries may cause breathing difficulties or send the victim into a coma.

Wild ivy is unable to stand on its own, but the main stem of a mature plant may be as much as 30 cm in diameter and standing out from a wall it can often be mistaken for a self-supporting tree. Both the Greeks and the Romans employed ivy extensively. In ancient Greece the priests presented newly married couples with a garland of ivy, to symbolise the fact that they were now bound together as closely as the ivy clasps its chosen tree. Ivy was also associated with Bacchus, the god of wine, who was often depicted crowned with vine and ivy leaves. *Hedera helix* is confined to temperate climates. In Asia Minor it is often quite dominant and in the Himalayas it is reported to be particularly common. The Revd C. A. Johns, writing at the beginning of the present century, pointed out that it was the Himalayan variety, with yellow berries, which was held in such esteem by the ancients, rather than our black-berried species.

Elder *Sambucus nigra*

The first time I was considered old enough to practise green cunning on my own followed the discovery that our beloved roses were covered with greenfly. I soon found an elder bush and brought back two baskets of elder leaves, which the little witch chopped and dropped into hot water. When cool, the liquid was decanted into a watering can, the roses

Ivy

were treated and the greenfly went. I have never used anything else on my roses since. Many insects seem to find elder leaves offensive and coachmen used to use elder twigs to make the whips for their horses – so there may have been a hint of kindness in their cruelty. When we went on family picnics just after the war the little witch always chose the site by keeping a sharp look-out for 'Judas bushes', because flies hated them. The name arose because legend has it that Judas hanged himself from an elder tree after betraying Christ, though there is no biblical evidence to support such an assertion.

In Cornwall the shrub is still held in great esteem as a medicinal plant and my great-grandmother made good use of all the remedies suggested by John Evelyn in his discourse on trees:

Elder

If the medicinal properties of the leaves, bark, berries etc. were thoroughly known, I cannot tell what our countrymen would ail, for which they might not find a remedy . . . either for sicknes or wound. The inner bark of elder applied to any burning takes out the fire immediately; that, or, in season, the buds boiled in water-grewel for a breakfast, has effected wonders in a fever; and the decoction is admirable to assuage inflammation. But an extract may be composed of the berries which is not only greatly efficacious to assist longevity but is a kind of catholico (universal preventive) against all infirmities whatever: and of the same berries is made an incomparable spirit, which, if drunk by itself, or mingled with wine, is not only an excellent drink, but admirable in dropsy. The ointment made with the young buds and leaves in May with butter is most sovereign for aches, shrunk sinews etc., and the flowers macerated in vinegar not only are of a grateful relish, but good to attenuate and cut raw and gross humours.

Elder ointment and elderflower tea still help to fortify the over-forties in our family and jars of elderberry wine bubble by the fireside every autumn. Tasting and looking like a dark, rich port – indeed unscrupulous nineteenth-century dealers were often found guilty of mixing the two – elderberry wine can sweat out a cold and alleviate gout. The berries, squeezed and mixed with honey, reduce fever and are a good source of vitamin C. The pale flowers, gathered fresh, make a pleasant sparkling white wine. If they are simmered gently in water, the decanted liquid is an excellent skin lotion.

Elder bushes grow quickly and the stems have a soft pith in the centre which can be easily scraped out. The little witch kept the pith in her first-aid box and many a painful household scald was soothed by elder. After the pith was removed, usually with a spoke from an old bicycle wheel, a hollow wooden pipe was left. With great skill she then used a tiny gimlet to make holes along it to produce a pleasant-sounding instrument. Thus she maintained an ancient tradition which can be seen in the botanical name for elder, *Sambucus*. This comes from the Latin word *sambuca*, which means 'pipe'. Pliny wrote that 'Countrymen believe that the most sonorous horns are made of elder which has grown where it never heard the cock crow.' Elder was for many years important to biologists, who held small specimens firmly between two pieces of the central pith so that they could slice them with a razor. Although more sophisticated techniques are now available, some botanists still prefer to use elder. A group once complained to me that elder was not so freely available as in former years and is also quite expensive. I smiled smugly as we went into the local wood, carefully pruned a couple of branches and ten minutes later had them ready for drying.

The Field Layer of Flowers

It is impossible to go into a wood and not find some plant growing. The idea that winter is the dead season is not true, it is merely less alive than spring. The reason many of our most beautiful flowers are at their best early in the year is that they have to get their flowering period over before the dominant trees come into leaf and block out the light. It is in March, April and May, therefore, that most of the useful herbs of the woodland are at their best.

Lesser Celandine *Ranunculus ficaria*
There are some charming names for this beautiful little plant. In Dorset it is known as 'spring messenger' and in Devon and Cornwall it is called 'bright eye'. The little witch, every inch a Cornish woman, had a favourite spring walk. 'Come 'ee dearie,' she would say, 'let we be off to look at broight oi.'

For the naturalist the lesser celandine is of historical interest. It is given an honourable mention in a herbal produced by William Turner in 1548. He called it figwort, whilst many other herbals called it pilewort. In fact, the two names are synonymous, since in classical Greece 'fig' was used as a euphemism for haemorrhoid. Sixteenth-century physicians called piles *ficus* and lesser celandine was used to treat this uncomfortable complaint. If you could afford it, the roots were boiled in wine. If you couldn't afford wine, then you used your own urine. It should be stressed, however, that the leaves contain chemicals which, although not lethal, can cause bleeding and inflammation, if taken internally. Their use is therefore not to be encouraged.

Another, less drastic, use was derived from the plant's physical characteristics. The underground root tubers resemble a cow's udder — as well as piles — and the

Lesser celandine

shiny, silky petals are the colour of butter, so our ancestors hung celandine in cowsheds to improve milk yield. It failed to work unless care was taken to collect the plant at the right time and in the right place, and it also required the correct spell:

I will pluck the figwort
With the fullness of sea and land
At the flow of the tide not the ebb.

The old botanical name for the species was *Chelidonium minus*, which became translated into 'lesser celandine' to distinguish it from *Chelidonium majus*, the greater celandine. It is now known that the two plants are not related, despite a slight, superficial resemblance. *Chelidon* is Greek for a swallow, and in ancient times it was thought that adult swallows bathed the eyes of their young with the dew from celandine leaves.

The little witch had one use for the plant which definitely works. She gathered the yellow petals and encouraged us all to rub our teeth with them, insisting that this was done in the days before anyone had ever heard of toothpaste. It certainly leaves a fresh taste in the mouth and I still occasionally clean my teeth with celandine petals when camping in the springtime.

Wood Anemone   *Anemone nemorosa*

The wood anemone dominates many British woods from April to June, though the odd flower can still be found in sheltered spots during August.

It is also very common on the Continent, especially in France, where it is called *l'herbe au vent*, or 'flower of the wind'. Similarly, the name anemone owes its origin to *anemos*, the Greek word for wind. You can easily see how the name came about if you sit quietly in the

shade of a wood and watch the delicate white flowers quivering when nudged by even the slightest of breezes. This trembling makes the wood anemone a difficult plant to photograph. The ancient Greeks had another name for the anemone. When Venus heard of the death of Adonis she was said to have wept so copiously that her tears changed into wood anemone flowers, thereafter called 'tears of Venus'.

All parts of the plant are quite poisonous and can even cause nasty blistering if crushed in the hand by a person with sensitive skin. The poison has been analysed and is called protoaneimonine. Culpepper, the knowledgeable, but often opinionated sixteenth-century herbalist, paid no heed to the poisonous reputation of the plant and strongly advised his patients to chew the roots of anemone, since 'it pergeth the head mightily and is therefore good for the lethargy. And when all is done let physicians prate what they please, all the pills in the dispensary purge not the head like hot things held in the mouth.'

Culpepper may well have cured the lethargy, but no doubt those of his patients that survived purging with wood anemone soon gave him the sack.

Wood Sorrel   *Oxalis acetosella*

Growing amongst the trunks of trees from April until July, wood sorrel is one of our prettiest woodland flowers. Despite its name, it is not related to sorrel or sheep's sorrel, which are members of the dock family. The biting taste is still enjoyed by country folk and its botanical name derives from the Greek word *oxys*, meaning sharp. This acid property has been exploited in several ways. Dairy farmers used to add wood sorrel to milk in order to curdle it for butter-making (though it should be noted that oxalic

Sorrel (**left**) and wood sorrel (**right**)

acid derivatives can be dangerous if consumed in large quantities). Also, oxalic crystals, made by pressing out the juice from wood sorrel and extracting the crystals by evaporation, used to be sold as a stain-remover under the name of 'essential salts of lemon'.

As early as the fifteenth century, wodesour, as it was called, was used in the preparation of a green sauce and cultivated in gardens for this purpose. Gerard noted that, 'When stamped and used for greene sauce, it is good for them that have sicke and feeble stomackes; for it strengtheneth the stomacke, procureth appetites, and of all sorrel sauces is the best, not only in virtue but also in the pleasantnesse of its taste'. For my own part, I can vouch for its excellence when baked in butter.

Thanks to its trifoliate leaves wood sorrel was regarded as a holy flower, the three leaves representing the Father, the Son and the Holy Ghost. Indeed, it is believed by many natural historians that St Patrick's flower was not the uncommon shamrock but the ubiquitous wood sorrel, which flowers freely throughout the British Isles and is usually in full leaf by St Patrick's Day, on 17 March.

William Turner in his herbal of 1568 mentions two vernacular names for wood sorrel: 'Alleluya', which may either refer to the fact that it is in full leaf at Easter or to the notion that its leaves recall the Trinity – or 'cuckoo flower', since it is in leaf at the time when the cuckoo arrives in Britain.

In the Hebrides 'Alleluya' was one of the constituents of a concoction made to cure scrofula, otherwise known as the King's evil. Geoffrey Grigson in his book *The Englishman's Flora* suggests that this could be the origin of the Buckinghamshire name for the plant, which is 'king finger'.

The plant itself is extremely sensitive and the leaves tend to close up at night or when rain is imminent. When dissected, the flower is revealed to be extremely regular, consisting of five sepals, five petals, ten stamens and five stigmas. Even the fruit is a five-angled capsule, so delicate that when it is ripe the slightest touch will cause it to explode and discharge the seeds within it over a considerable distance.

An unusual feature of wood sorrel is the presence of what are called cleistogamic buds. These never open, but the seeds still develop within them and, what is more, are actually viable. There is much more to this plant, with its small white flowers streaked with purple and red, than meets the eye. But then that is true of all our flora.

Ramsons *Allium ursinum*

Because of its strong onion-like smell this plant is also known as wild garlic. It has been used in cooking since Anglo-Saxon times and Gerard mentions that a fish sauce could be made from its leaves. However, he only recommended the sauce to those who were 'of strong constitution and labouring men'. Personally, I find the taste overpowering, but in the Middle Ages food was often past its best when eaten and strong herbs could often disguise rotting meat and fish.

My great-grandmother had several uses for 'stinking onions', as she called the plant. Like me, she savoured the smell of garlic, but found the taste far too strong. She got round this by warming the plates, then wiping them with the leaves of ramsons before serving the meal, a method I still find useful when eating steak. She also made garlic ointment by mixing chopped boiled ramsons with the agar jelly she made from seaweed. Rubbed vigorously into aching joints or

Ramsons

bruised limbs, the ointment generates a warm, tingling sensation and brings quick relief.

The name 'ramsons' is descended from *hramsa*, the Old English word for the plant, and has found its way into many place-names, such as Ramsey and possibly Ramsbottom in Lancashire – although some argue that the significance of the latter is that breeding rams were penned there. Many damp wooded valleys abound with ramsons and the heady scent is all too obvious after rain or when the juicy leaves are trodden on. The leaves are about 15 cm long, dark green and very succulent. Inside the leaves is the triangular flower stem called the scape, and at the top of the scape are borne about a dozen white, star-shaped flowers. The whole plant grows from an underground bulb, which clearly reveals the close relationship of ramsons to the onion family. The onion is named *Allium cepa* and ramsons *Allium ursinum*, *allium* being the Latin for garlic, while *ursinum* means 'pertaining to bears'. Could it be that the bears of Britain ate ramsons before they were hunted to extinction? In Wiltshire ramsons are still occasionally called badger flowers, so perhaps there are gourmets in the animal kingdom that enjoy a well-flavoured meal.

Lily of the valley

Lily of the Valley *Convallaria majalis*
The lily of the valley has a long history of medical use and was known to the ancient Greeks as a cure for heart disease and dropsy. However, great care is needed when the plant is used for medicinal purposes and the correct dose is crucial, since the whole plant contains poisonous glycosides, the effects of which are similar to digitalis in foxglove, though a little less drastic in their action. Glycosides cause sickness, vomiting, and pains in the head and digestive tract. In the initial stages following administration they slow the pulse, but if the dose is too large the heartbeat eventually accelerates, loses its rhythm and finally fails altogether. Gerard suggested small doses for the treatment of fevers, weak memory and eye trouble. He also suggested that if the flowers of lily of the valley were placed in a glass vessel and buried in an ant hill the liquid resulting could be used as a cure for gout. And in more modern times, during the First World War, lily of the valley was used successfully to treat those suffering from the horrific after-effects of gassing.

Some of the vernacular names such as

'ladder to heaven' and 'fairies' bells' indicate the poisonous nature of the plant. Two other names, 'innocents' and 'lady's tears' are associated with a German (probably Saxon) folk tale which relates how the tears of either the Virgin Mary at the foot of the cross or Mary Magdalene at the deserted tomb were changed into lily of the valley.

Lily of the valley is a perennial plant, sprouting each spring from a branched underground stem, or rhizome, with leaves 10–20 cm high arising in pairs. One of its favourite habitats is limestone woodland, but in years gone by so many blossoms were picked during May and June that the plant is no longer common in the wild. Lily of the valley is perhaps the sweetest-smelling of all our woodland flowers. The pendulous white blossoms are pollinated by insects attracted by the delicate perfume. The berries, which are green at first and ripen to red in August, are permeated with poison.

Cuckoo Pint (Wild Arum)

Arum maculatum

When I first set out to discover something about this strange plant I soon found myself completely confused. I kept coming across names like lords and ladies, wake robbin, parson-in-his-pulpit, friar's cowl, red-hot poker and Jack-by-the-hedge. In fact, it is the proud possessor of over fifty vernacular names.

Cuckoo pint grows in damp, shady places and is common throughout Britain. It is a perennial plant and, like lily of the valley, survives the winter as a rhizome which in spring provides food for the developing aerial parts.

The structure normally referred to as a flower is actually a much more complex structure called an inflorescence. This is made up of a yellow-green spathe or bract surrounding a purple-coloured column

called a spadix, which resembles an erect male organ and accounts for several of the plant's more colourful vernacular names. The true flowers are situated on the lower part of the spadix, where they peep out of a cavity in the bract. The plant is pollinated by small flies of the midge type attracted by the smell of nectar emanating from it, which resembles the stench of rotting meat. The flies are allowed to enter by a small chamber in the bract, the entrance being protected by small, downward-pointing hairs. This arrangement permits free entry, but prevents exit. Once it has gained entry, the fly can only

Cuckoo pint or wild arum

move downwards until it encounters a cluster of male flowers. Below them is yet another row of downward-sloping hairs, leading to the female flowers. Here the fly finds itself trapped – but at last it gets its meal, since the nectar which attracted it to the plant in the first place is produced by the bases of the female flowers. If a thermometer is inserted into the plant at this stage it shows quite a significant rise in temperature – but what the precise function of this is nobody knows.

The female flowers mature before the male flowers, but if the fly has previously been buzzing about within the confines of another cuckoo pint plant it will still have pollen grains adhering to its body and these may well be deposited on the stigma of its present captor. Once the female flowers have been pollinated, the male flowers become mature, produce pollen and the downward-sloping hairs wither, allowing the fly to escape if it has not died during its imprisonment. As it journeys upwards the insect becomes dusted with pollen. It then flies away in search of more nectar produced by another cuckoo pint with unfertilised female flowers.

Cuckoo pint is usually in flower from mid-April until the end of May. Its autumnal aspect is very different. We look in vain for blossom and foliage. Instead we find a green stem with a cluster of some twelve scarlet berries at the top. These are unpleasant to the palate and, despite the fact that many birds eat them, to human beings they are extremely poisonous. It was suggested in a seventeenth-century herbal that the berries should be served to a 'sawcey guest' for 'within a while after the taking thereof, it will so burn and pinch his mouth and throat that he shall not be able to eat any more or scarce to speak for pain'. Which seems a somewhat drastic

way to get rid of a bore!

There is an old country belief that thrushes dig up and eat the roots of arum, but since thrushes are not usually root eaters this scrap of folklore must be treated with a certain amount of caution.

The name *Arum* is of Greek origin and may derive from *aur*, which in the ancient Hebrew and Egyptian languages was the root-word for fire, while *maculatum* is the Latin for spotted – since the plant sometimes, though not always, has spotted leaves.

Should the rhizome be eaten raw the taste is most unpleasant, but in times gone by it was carefully dried and found not only to be harmless but so nutritious that it could be used as a substitute for cornflour. It was at one time sold under the name of Portland sago. Wild arum was also used extensively in the preparation of powder for wigs and was employed as a starch for the maintenance of the elaborate ruffs that graced the necks of our Elizabethan forebears. As a result it was also known as starchwort, though the name is now obsolete.

St John's Wort *Hypericum* spp

To the modern naturalist St John's wort can be confusing, since there are at least nine species, some of which hybridise, so it is not always easy to identify the individual varieties. To our ancestors such minor distinctions were probably not apparent and on Midsummer Night they were quite likely to return from their wanderings in the fields laden with a mixture of species, including tutsan (*Hypericum androsaemum*) and Rose of Sharon (*Hypericum calycinum*). The old name St John's wort came from the fact that the flowers were used to keep evil spirits at bay on St John's Eve – and one of the plant's old names was *Fuga daemonium* or Devil's bane. Its effect was supposed to be even more potent if the flower was still shining with early morning dew when it was picked. In Saxony it was customary for unmarried girls to seek the plant on St John's Eve. There was even a verse to encourage them:

Thou silver glow worm, oh lend me thy
  light,
I must gather the mystic St John's wort
  tonight;
The wonderful herb, whose leaf will
  decide
If the coming year shall make me a bride.

Each girl, after finding and picking the plant, would keep it in her bedroom. The fate of the flower was supposed to determine whether or not she would marry within the next twelve months. If the flower died quickly, the girl had no marriage prospects; the longer the flower lasted, the better her chances were.

The appearance of the flower may well have been the reason for its association with John the Baptist, who lost his head at the whim of Herod's daughter, since its yellow flowers with reddish sap and stamens were thought to represent bloodstains. One species, perforate St John's wort (*Hypericum perforatum*), has a line of black dots on the leaves that look like perforations when they are held up to the light. To the medieval mind these represented stab wounds, so it was regarded as an ideal plant for treating wounds or abrasions − and the fact that it was a holy plant rendered it an even more potent remedy.

Primrose *Primula vulgaris*
Before the Wildlife and Countryside Act came to their rescue the primroses in the remaining woodlands of Britain were declining at an alarming rate. I must admit that my great-grandmother and I did our

Primrose

share of damage, since we frequently collected bunches to decorate the old house and the family grave in the churchyard up on the hill. Early on an April morning, often with a chill wind blowing off a grey sea, we used to make our way into the shelter of our sweet-smelling woodland where primroses bloomed under my ash tree. The little witch was a strange combination of the superstitious and the practical. Nothing would persuade her to carry fewer than thirteen primroses into the house for fear that this would stop the hens from laying. Once we had gathered enough flowers she would set about mixing chopped primroses into boiling mutton fat to produce a soothing ointment which healed cuts and bruises as fast as anything I have used since. Anyone with a sore eye quickly had it bathed with warm water in which dried primrose leaves had been soaked for five minutes.

By May Day primroses are at their best and, as witches were supposed to be at their most active at this time, it is not surprising that such a pretty flower should have been employed to drive the old hags away. What is surprising is that poets associate the primrose with sadness. In

Greek mythology Paralisos, son of the goddess Flora, died of grief when he lost his lover and was reincarnated as a primrose. In *Cymbeline* Shakespeare continues the theme, and when Arviragus thinks Imogen is dead he speaks these lines:

I'll sweeten thy sad grave; thou shalt
not lack
The flower that's like thy face, pale
primrose, nor
The azur'd harebell like thy veins, no,
nor
The leaf of eglantine.

Bluebell *Endymion non-scriptus*

To my mind one of the best features of the British countryside is a bluebell wood in late spring or early summer. Indeed, to most people the beauty of the bluebell is so delicate that it comes as a surprise to learn that it once had a mundane, though practical use for our ancestors. They produced a thick, pungent glue by grinding the bulbs to a pulp, then heating the mush to drive off the excess water. When squeezed, the stems too produce a glue – so strong that fletchers used it to cement the feathered flights to the shafts of arrows. I still have a football scrapbook dated 1949 which has pictures stuck in with bluebell glue.

In its time the bluebell has been given many vernacular names, including 'pride of the woods', 'adder flower', 'crow flower' and 'crow toes', and in some parts of England it gloried in the delightful name of 'Granfer Grigglesticks'. In Scotland, where 'bluebell' is the name used for the harebell or *Campanula rotundifolia*, it is called wild hyacinth.

The curious botanical name of the bluebell has a scientific explanation. Before the appellation *Endymion* was adopted the plant was known as *Hyacinthus non-scriptus*. The petals of members of the true

hyacinth family appear to be inscribed with the initials A. L. If you look at the flowers of a hyacinth closely, these initials can be seen quite clearly, just as if someone has signed the petals. But on the petals of the bluebell (which is a member of the lily family) there are no initials, which is why it is called *non-scriptus*, 'without writing'.

The Ground Layer

Growing close to the woodland floor are plants such as liverworts and mosses which are shaded out by the taller plants and are therefore obliged to get their flowering period over in the winter months when, although there is not much light, at least they do not have to fight for it.

It is difficult to assign a precise position to the ferns, however in their early stages all species are small enough to be included in the ground layer. Whilst in the mature stage – especially of species such as bracken and male fern – they may form an important part of the field layer.

Orpine *Sedum telephium*

Orpine is an attractive member of the stonecrop family. Unlike other stonecrops, it has a spotted stem and much broader leaves. The flowering period is from July to the end of September and favourite habitats are damp woods and waste places. Both the size of the plant itself and the colour of its flowers vary a great deal, but the sepals are reddish and the petals white, green or yellow.

Orpine has a great tradition in the history of medicine and has also been closely associated with midsummer festivals, as many of its vernacular names clearly show. These include 'healing leaf', 'livelong' and 'midsummer men'. The leaves were gathered and included in

Bluebell

Orpine

posies used to decorate houses or to burn on a purifying fire to drive away evil. They can also be boiled and eaten. The plant keeps well and is easily grown in pots. At one time it was used to detect the presence of witches, for orpine was supposed to wither whenever anyone associated with the Devil entered the room. Other, more practical, uses included bringing down fever, curing sterility and reducing excessive menstrual bleeding, and Culpepper recommended it as a cure for cancerous growths. The convoluted shape of the root tubers ensured that orpine was regarded as a signature plant and it was thought that they could combat the King's evil. Even today country folk still place orpine on cuts and abrasions, since – like the dog rose – it is an efficient astringent.

Ferns
The folklore surrounding ferns concerns not so much plants themselves as the spores, which are so tiny that they can barely be seen by the naked eye. It was therefore believed that ferns flowered only once a year, at midnight on Midsummer Eve, the flowers remaining invisible for the rest of the year, and that some vital force gave rise to new ferns. It followed that if the secret of this invisible force could be discovered, then it would be possible to travel into the spirit world and so stand a better chance of avoiding being seen by the Devil or his minions. This arduous journey was best attempted by young men, who ate fern plants off gleaming platters before setting off full of the spirit of chivalry, while the maidens dutifully waited, virtue intact, for their return. All the eager maidens could do was to dream wistfully of marriage – but if only they could discover who they were going to marry, it might ease their heartache. Sometimes the bolder ones were willing to chance eating a few fern 'seeds' themselves and venture into the village churchyard. In the churchyard, the first eligible male they saw was destined to be their future husband.

Ferns also had more down to earth applications. They were used for cattle food and as bedding. Potash of quite good quality could be produced from ferns too and this was at one time a profitable woodland industry.

Bracken *Pteridium aquilinum*
Many of Britain's woodlands and hillsides are clothed with bracken, which has a world-wide distribution. Its tough green fronds are ignored by most grazing animals and are poisonous when fresh. In its early stages, however, bracken consists of a tiny, delicate, heart-shaped prothallus, which is very much part of the field layer. Bracken grows from a stout underground rhizome packed with carbohydrate. At one time this was dug up and ground into a flour used to make bread.

While writing this book I tasted bracken bread for the first time and found it rather like oat cake, if a trifle bitter.

My great-grandmother used the fronds to produce a delicate brown dye, and told me that in the local slate quarries bracken was collected and used as protective packing. She also used bracken on her compost heap and maintained that bracken and seaweed made the best possible fertiliser.

When camping in autumn, dry dead bracken makes a far better bed than any inflatable mattress or tent lining I have yet discovered.

Maidenhair Spleenwort

Asplenium trichomanes

Although more often found on walls, in some places this fern grows in the ground layer, particularly on damp sloping banks. Here it occurs with liverworts and the two were often used together by the old herbalists in the treatment of the liver and spleen. The little witch was convinced that maidenhair spleenwort was a hair tonic and regularly used it herself. It has to be said that, with her splendid head of hair, she was an excellent advertise-

Maidenhair spleenwort

ment for it. From the field layer she also collected mosses, which she mixed with fumitory to keep away insects. Distant relatives got used to receiving presents through the post packed in moss and were surprised how well they travelled. But then the little witch was always full of surprises.

7 Plants of the Fields and Hedgerows

Hedgerow Plants

The little witch was a great hedgerow hunter. Recently, I retraced one of her old routes through a farm, thankfully still owned by the same family, with a grandson who is an enthusiastic farmer. Such farmers are usually just as keen on conservation as the most ardent naturalist, if not keener. I had not been walking long when the smell of wood smoke drifted across a field and there I came upon a hedger at work, deftly bending and twisting the trees and bushes to his will and cutting out and burning dead or unwanted branches which the animals could break through. He called me over to join him and it turned out that his grandfather and the little witch had often passed the time of day. I enjoyed a potato roasted among the ashes of his fire and a slice of his blackberry pie before taking a look at the hedge, which I found included several elms, sycamore, hawthorn, sloe, crab apple and a solitary spindle.

Elm *Ulmus* spp
Culpepper wrote of elm thus:

It is a cold and Saturnine plant. The leaves thereof bruised and applied, heal green wounds being bound thereon with its own bark. The leaves or the bark used with vinegar, cures scurf and leprosy very effectually. The decoction of the leaves, bark or roots, being bathed, heals broken bones. The water that is found in the bladders on the leaves, while it is fresh, is very effectual to cleanse the skin and make it fair; and if cloths be often wet therein and applied to the ruptures of children it healeth them if they be well bound up with a truss. The said water juice taken therein, killeth and driveth forth all manner of worms in the belly, stomach, and maw; and gargled in the mouth or the root chewed, fasteneth loose teeth, and helps to keep them from putrefaction; and being drank, is good for those that spit blood, helpeth to remove cramps or convulsions, gout, sciatica, pains in the joints, applied outwardly or inwardly and is also good for those that are bursten, or have any inward bruise. The root boiled well in vinegar, beaten afterwards, and made into an ointment with hog's suet or oil of trotters, is a most excellent remedy for scabs or itch in young or old; the places also bathed or washed with decoctions doth the same; it also helpeth all sorts of filthy old putrid sores or cankers whatsoever. In the roots of this herb lieth the chief effect of the remedies aforesaid. The distilled water of the roots and leaves together is very profitable to cleanse the skin of the face, or other parts, from any morphew, spots or blemishes therein and make it clear.

Culpepper may well have been right about the value of many of the uses he suggests for elm, since the tree is related to the nettle, which has proved effective for treating painful joints, and people with sensitive skin have been known to blister after handling elm leaves.

Many varieties of elm grow in Britain. However, the medical virtues of all the species are similar and the two most commonly encountered are the wych elm (*Ulmus glabra*) and the commoner English elm (*Ulmus procera*), which has been devastated by Dutch elm disease in recent years.

The wych elm has more spreading branches than the common elm and the ends of the branches often droop, giving the appearance from a distance of a weeping willow. The leaves are much rougher than those of the common elm. They are also much larger and are narrow at the base, but widen towards the apex.

Elms blossom early in spring, often as early as February, but the flowers are at their best in March. When the male flowers are ripe, if struck by the rays of the sun, the tree seems to glow with a purple light. When they have withered, the female flowers come into their own and the tint changes to brown. The seed vessels, which blow about in the wind, look like small, flat green leaves, but if viewed against the light the seed can be seen in the centre. The common elm has seed vessels that are smaller than those of the wych elm and my great-grandmother used to collect them when they were green and chop them up with dilute vinegar to produce a mouthwash which was most effective against ulcers.

Many species of elm spread by means of suckers and their seeds tend to be infertile. The wych elm, however, never has suckers and its seed is fertile. As a result it is the most widespread of the British elms. Its timber is excellent and was once used for building small boats and, especially in Wales, for the manufacture of bows and arrows. Given its natural durability, elm timber was invaluable for making the keels of warships and the groynes of harbours, since it contains a chemical which reacts with seawater and prevents decay. The interlocking grain also made it an ideal choice for wooden wheels, chair seats, mallet heads and other items likely to be subject to abrasion. In the early days of piped water the supply was carried by hollowed-out elm trunks dovetailed together. Lift and force pumps were also made out of elm before the smelting of metals improved and non-rusting alloys were discovered.

The little witch respected elm medicine but hated the tree, which she insisted was sly enough to drop heavy boughs on the unsuspecting. There is some truth in her accusation, since elms tend to rot from the inside then tumble without warning. Elm, unlike sycamore, is therefore a poor choice for a shelter belt.

Sycamore *Acer pseudoplatanus*
The sycamore can be recognised in winter and spring by its green buds, and in summer by its five-pointed leaves shaped like a splay-fingered hand. The wood is creamy white in colour, hard, strong and smooth. It is used for furniture, dance floors, ladders, dairy utensils, spoons and for rollers in textile machinery. Occasionally trees show a beautiful curly grain or ripple figuration and these are very much in demand for veneers and by violin makers. This peculiarity is almost certainly genetic, but nobody seems ever to have tried to breed wavy-grained sycamores. The species grows well in Italy and has been associated with violin-making for centuries. Sycamore is used for the sides, back and stock, but the sounding board is always fashioned from a conifer, usually Norway spruce. The

Sycamore

reason for this combination is largely structural, the sycamore being readily carved, but it is also said that it vibrates at a different musical interval from the spruce, giving a pleasing combination of harmonics.

Sycamore wood is used in the textile trade because its creamy surface wears well and is free from any tendency to stain fabrics. It was used mainly for rollers, some being turned from a single log, others built up from several pieces. Sycamore is a splendid timber for turning, since it does not splinter when rotating, and it can be stained grey with iron salts to produce 'harewood', a popular furniture timber.

The little witch made the transition from Cornwall to the North with remarkable ease, but occasionally the harsher climate annoyed her – especially when Easter was early and she could not bake her revel buns. For these she needed fresh sycamore leaves, which she wrapped around each tiny bun before putting it in the oven and which made the buns sweet and particularly tasty.

Some botanists believe that the sycamore was introduced into Britain from the Balkans in the sixteenth century, whilst others argue that if it is not native then it was introduced by the Romans as a timber tree. There is no doubt that the sycamore was planted as a shelter tree and also that it was quite possible to confuse it with the native field maple (*Acer campestris*) which, although more bush-like, can grow to a height of over 20 metres and produce useful timber. The leaves of field maple are smaller and more deeply cleft; and the fruit is a different shape from that of sycamore. The seeds of both species are carried between two 'wings' that spin in the air, which is why children call them 'helicopters'. The wings of field maple are horizontal whilst

The leaves and keys of sycamore (**top**) and field maple (**bottom**)

those of sycamore keys are sharply angled. In hedgerows field maple is probably more common than sycamore, except in Scotland, where field maple is less common, especially in the North.

Hawthorn *Crataegus* spp

'Ne'er cast a clout till may is out' the little witch used to insist, like many a mother and grandmother, and we waited till June before we ventured out without a 'ganzy', or pullover, on. Others hold that the true meaning of the proverb is that one should not shed winter clothes until the hawthorn blossom is in flower or has faded.

The ancient Greeks regarded this time of year as inauspicious – a season when purification ceremonies were to be enacted and lovemaking was to be avoided. In parts of Spain, particularly the northeast, such customs are still observed and it is held to be unlucky to wear new clothes.

In pre-Christian Britain May marriages were viewed as impropitious and may even have been forbidden. In the Welsh romance of Kilhwych and Olwen the father of the hopeful bride appears before the lovers as Giant Hawthorn in an attempt to prevent the imminent nuptials – clear evidence of the taboo against marriage in the month dominated by the may tree. Across the sea in Ireland hawthorn is called 'hagthorn' and it is considered unlucky to destroy a may tree.

Originally, the first of May was celebrated as a spring festival, with rites and rituals designed to promote the fertility of both humans and animals. On the first day of May Celtic country folk released their cattle into the fields and on the first of October brought them in again, both dates being closely associated with witchcraft. As agriculture began to include an arable element, May Day evolved into a celebration of the goddess Flora, the May Queen, whose maidens gathered hawthorn blossom, with its sweet, faintly sexual aroma, and danced round the maypole, itself a phallic symbol. Since may blossom could not be picked without risking the vengeance of the witches, rowan twigs were gathered too, to ward off their spells. Jack-in-Green, or the Green Man, often seen in inn signs near village greens and like the May Queen a powerful fertility figure, also played a prominent part in May festivals.

The may tree no longer blossoms on the first of May, as it did in former times – since, when the New Style calendar was introduced in 1752, 1 May became 13 May and hawthorn is rarely in flower before that date.

Two species of hawthorn are native to Britain. The most common is *Crataegus monogyna*, which has white flowers, fairly deep cut leaves and deep-red haws, each containing a single seed. The second species, Midland hawthorn (now called *Crataegus oxyacanthoides*, though formerly known as *Crataegus laevigata*), has haws that are rounder and brighter than those of common hawthorn and each contains two seeds. It interbreeds freely with common hawthorn, and the flowers of Midland hawthorn and of the hybrids may be either single or double, varying from pure white to pink or scarlet. The leaves have round lobes and are less deeply cut. Common hawthorn is a true pioneer species, occurring throughout the British Isles, especially in areas that have been recently cleared. Midland hawthorn, on the other hand, prefers deeper, heavier and richer soil and, despite its name, is found mainly in the south of England. It is also more tolerant of shade.

Hawthorn is a tree of great importance to many animals, the blossom being essential for the survival of early hoverflies and several species of beetle, including the hawthorn shield bug. No less dependent on hawthorn are the autumn birds. Blackbirds, mistle thrushes, fieldfares and redwings all find the rich red haws irresistible. The insects repay the tree by pollinating its flowers, whilst the birds pass the seeds through their digestive tract and often void them many miles from the parent. Carefully controlled research has shown that seeds which have made the journey through the birds' digestive system germinate more successfully than ones which have not.

Those who leave the haws to the birds may well have missed one of nature's

Hawthorn

greatest treats, for hawthorn wine is surprisingly like pink champagne. It can be made by the usual home-brew method, but the maturing wine needs to be stored carefully. We once heard a series of explosions in the night and came downstairs to discover that the corks on our hawthorn wine had blown and the kitchen ceiling was dripping with sticky pink liquid.

Old Man's Beard *Clematis vitalba*
Old man's beard or 'traveller's joy' is often found climbing and winding around hawthorn and other trees and bushes, especially those growing in hedgerows and at the edges of woods on chalk and limestone. Because it binds so tightly it was once included, symbolically, in bridal bouquets. It was also used in attempts to cure baldness, no doubt because of the hairy appearance of the autumn seed heads.

Blackthorn (Sloe) *Prunus spinosa*
According to Richard Mabey the ancient Hibernian club or 'tranquilliser' called a shillelagh was made from a stout bough of blackthorn. Because the flowers usually appear before the leaves, sloe blossom is among the earliest signs of spring, bringing to the countryside a promise of kinder weather, even though a chill north wind may be driving across the land.

Blackthorn spreads not only through the distribution of the seed within its edible fruit but also by putting out suckers, which soon produce a substantial thicket from a single tree. The thorns deter grazing animals from nibbling the shoots and the buds are surprisingly resistant to wind and salt spray. As a result, blackthorn is an ideal hedging plant. Its leaves were once gathered, dried and brewed like tea – and when my great-grandmother mixed blackthorn leaves with her wartime tea ration no one noticed the difference. She always said that in the early days of tea drinking in Britain unscrupulous dealers made their fortunes by selling such a mixture.

The sloes themselves can be made into sloe wine or sloe gin, a practice returning to favour now that home brewing is so popular. The French name for blackthorn is mère-de-bois, no doubt relating to its tendency to grow around woodland margins and to throw up new shoots from suckers, and in France unripe sloes were once sometimes used as a substitute for olives.

In England juice squeezed out of the unripe fruit was at one time sold under the trade name of German acacia and used to mark linen. It made a first-class laundry marker since the mark would not wash out. The bark of the tree was used in the manufacture of ink, and if it is boiled in alkali an attractive yellow dye is produced. As well as polished clubs, blackthorn wood served to make smooth, stylish walking sticks, hay rakes and flails for threshing corn.

Crab Apple *Malus sylvestris*
The crab apple is a shrubby deciduous tree, seldom growing higher than 15 metres, with delightful pink blossom that opens during May. The crab apples, which are green at first then turn a delicious shiny red, generally ripen by late September. They are usually far too bitter to eat raw but make a splendid wine, and each year the little witch could be relied on to produce a store of crab apple jelly by boiling the fruit with honey and a small quantity of her seaweed agar. She also mixed crab apples with blackberries to make excellent pies.

Crab apple wood was at one time used for the heads of mallets and as firewood it has no peer. After cultivated apples had

Old man's beard or traveller's joy

Crab apple

a strong purgative. The smooth wood was used to make spindles (hence the shrub's name), keys for virginals and bows for violas.

The Hedgerow Understorey

Like a woodland, a mature hedgerow has a well-developed group of understorey plants supported or shaded by the larger plants. Three of the most common are honeysuckle, dog rose and blackberry.

Honeysuckle *Lonicera periclymenum*

> His companions were timber dealers, yeomen, farmers, villagers and others; mostly woodland men who on that account could afford to be curious in their walking sticks, which consequently exhibited various monstrosities of vegetation, the chief being cork screw shapes in black and white thorn, brought to that pattern by the slow torture of an encircling woodbine during their growth.

Thus wrote Thomas Hardy in *The Woodlanders* in 1887 of a walking stick fashioned by the twining action of honeysuckle. My great-grandmother had such a stick for many years. They were called 'barley sugar sticks' and were a great favourite with music hall comedians.

From honeysuckle leaves my great-grandmother made a mouthwash for ulcers and 'dirty tongues'. She also mixed the leaves and blossoms (half a cup of each) in a pint of warm water and drank a small glass of this mixture before meals, saying that it improved both appetite and digestion. In the past honeysuckle had a great many medicinal uses and the bark was thought to cure dropsy. The plant also had a place in folklore and it was believed that it could avert evil spirits,

supplanted them as fruit, the once valuable crab trees were plundered for firewood, but in the ancient wildwoods they were principally a source of autumn food.

Spindle *Euonymus europaeus*

The spindle is not a common wild shrub in Britain, especially in the North. It grows best on chalk and limestone. The only specimen in our village was part of a hedge, where more than likely it was planted, though it is just possible that it could have been bird-sown. My own memory of our spindle is not a pleasant one. When school started after the long hot summer of 1947 there was the inevitable outbreak of headlice. The thought of nits in her bairns' hair appalled the little witch. Off we went up 'mucky lonning', a long lane winding high up on to a limestone bluff, and from the spindle tree we collected the round orange fruits she called 'louseberries'. These were dried in a baking tin and when brittle were ground into a powder which was rubbed into our hair to get rid of the lice.

Spindle berries had their place in medical folklore and were administered as

which were particularly active on May Day.

Honeysuckle is a deciduous woody climber which twines clockwise. Often as early as December it produces opposite pairs of smooth oval leaves, either with very short stalks or no stalks at all, and the heady perfume of its flowers pervades the countryside from June until September. The flowers are about 5 cm long and occur in clusters at the ends of shoots arising from the main stem. The corolla (the collective name for the petals) is shaped like a long trumpet, with the mouth divided into lobes. Within it are five stamens, which are so long that they project beyond the corolla, and in the centre of the stamens is a single style. The flowers may be white, cream, yellow or even pink and have a great attraction for bees. After pollination the colour deepens.

The name *Lonicera* was given to the

Honeysuckle

plant in honour of Lonceir, a sixteenth-century German botanist, while *periclymenum* is of Greek origin and means 'to wind about'. The word 'honeysuckle' itself refers to the custom of country children, who suck the flowers in order to extract the rich, sweet nectar. It also possesses a great number of evocative vernacular names, including 'bindweed', 'bondweed', 'honeybind', 'trumpet flower', 'trombone plant', 'bugle blooms' and 'evening pride'. Probably the most widely used of these local names is 'woodbine', but my own favourite is 'lamps of scent'.

Dog Rose *Rosa canina*

In China rose petals are often eaten as a vegetable. This is hardly surprising as much of the plant is rich in vitamins, especially vitamin C. Rose hip syrup is still available commercially, but can be made easily enough by boiling 2¼ lb (1 kg) of hips, after removing the hairy seeds, with the same weight of sugar in a pint (0.6 litre) of water and straining off after simmering for twenty minutes. Culpepper's sixteenth-century herbal notes that the hips 'maketh most pleasante meates and banqueting dishes, and tartes and such likes'.

In medicine the rose has been put to a variety of uses. The petals are quite an effective astringent and the leaves were used to cure St Anthony's fire and the 'French pox'. The leaves and petals of garden roses can be used for herbal treatments, but the apothecaries eagerly sought a feature of the wild rose that seldom occurs in domesticated strains, namely 'bedeguars' or robin's pincushion galls. Galls are formed when an insect lays an egg or eggs in the bud of a plant. In the case of the dog rose the offending insect is a tiny wasp called *Rhodites rosae*. The plant defends itself by enveloping the part

Dog rose

attacked with a protective tissue, thus sealing off the grubs that develop from the eggs and preventing damage to other parts. Eventually the mature insects burrow their way out of the gall before mating. The old apothecaries gathered robin's pincushions, which are red and spiky, and either used them whole to staunch blood or cut them in half to get at the grubs, which they dried and used as a sort of signature brew to rid their patients of parasitic worms. The grubs were also hung round the neck in an attempt to cure whooping cough and placed under the pillow to cure insomnia.

The word 'dog' when applied to a plant usually indicates that it is common or at least was so at some time in the past – for example dog violet, dog's mercury and dog daisy. In the case of the dog rose, however, two other explanations have been put forward. It was believed that an extract made from the roots could ease the pain of a dog bite, an idea dating back at least to the time of the Greek writer Theophrastus, while Pliny reported that a soldier from the Praetorian guard used the root to cure himself of hydrophobia. Some authorities argue that 'dog' may be derived from 'dag', since the thorns are shaped like daggers.

The dog rose does not produce any nectar, but small flies and flying beetles land in the centre of the flower and gorge themselves on the rich supply of pollen. Their constant coming and going guarantees efficient pollination. Once the ovule has been fertilised by the pollen grain cell, the development of the plant departs from the standard botanical pattern. The hip is not a true fruit, but is formed when the end of the stalk swells up and surrounds the real fruits, which are hairy. The hip is therefore termed a false fruit. The hips are eaten by birds and mammals and, whilst the attractive red

pulp is easily digested, the true fruits within it pass through the gut relatively unharmed and so are spread far and wide. The flowering period is short, but the dog rose is always in bloom around Midsummer Eve – a fact which, together with its delicate perfume, simple beauty and medicinal properties ensured its place among the plants believed to drive out evil.

Blackberry *Rubus fruticosus* agg
The blackberry – also, almost everywhere in Britain, called bramble – belongs to the Rosaceae family, as is revealed by the shapely white flowers which are similar to those of the dog rose, though much smaller. There are hundreds of varieties of bramble, and blackberries were an important part of man's autumnal diet in the days before recorded history. Near Walton-on-the-Naze on the Essex coast in 1911 the body of a Neolithic man was discovered well preserved in clay, and blackberry seeds were found among the contents of his stomach.

The blackberry was much more than an autumn delicacy. In the Highlands of Scotland, for example, the leaves were used to treat burns, swellings and septic wounds. Blackberry was considered to be effective for the treatment of gout and rheumatism and the fibres taken from the centre of the stem were employed as twine. Such a useful plant was bound to gather around itself a blanket of superstition. Again in Scotland, ivy, rowan and blackberry were twined around each other and used to drive away evil. Far away in Cornwall nine bramble leaves were picked and purified by water before being placed on a wound. As the wound was treated, the following verse was recited three times over each leaf:

There came three angels out of the east,

One brought fire and two brought frost.
Out fire and in frost
In the name of the Father, Son and Holy
  Ghost.

The little witch stuck by this and also by another Cornish belief – that at Michaelmas the Devil urinated on all blackberry fruits. She steadfastly refused to pick a single berry after 29 September. Up till Michaelmas, however, her cottage hummed with activity as blackberries were bottled, jellied, jammed and fermented. Her recipes were delicious – and of course cost next to nothing.

Recipe for bramble jelly

Ingredients: 3 lb (1½ kg) blackberries, 3 lb (1½ kg) sugar. Put the fruit and sugar into a pan and add 3 pints (1½ litres) of water. Bring to the boil and simmer for half an hour. Strain the juice through a jelly bag and add a few drops of citric acid or the juice of a lemon. Bottle and cool.

Recipe for blackberry jam

The ingredients are the same as for bramble jelly, but the whole of the fruit goes into the jam, not only the juice. It need not, therefore, be strained. Occasionally the jam will not set, so I usually add ½ lb (250 g) of chopped crab apples. This not only improves the setting qualities, but adds considerably to the taste. The little witch also used seaweed agar to help setting.

Recipe for blackberry wine

This is a fine full, sweet red wine, but to my mind its greatest attribute is the wonderful colour. Whenever a bottle of my great-grandmother's blackberry wine was opened, we got out our crystal glasses. The colour of the wine, she said, deserved the best.

To make blackberry wine, take 4 lb (2 kg) of blackberries, 3 lb (1½ kg) of sugar and some yeast. First, wash the blackberries carefully and, after weighing them, place them in a plastic bucket and crush them with a wooden rolling pin. No metal objects should be used. Bring to the boil 8 pints (4½ litres) of water, pour it over the crushed fruit and stir regularly until the temperature falls to about 65°C. Leave for a few hours before adding the yeast (a little yeast nutrient is also beneficial at this stage), then cover the bucket with a plastic bag secured with a large elastic band or with string. The cover should be removed each day and the contents stirred. On the fifth day, squeeze the contents of the bucket through a muslin wine bag, letting the juice fall on to the sugar in another bucket. Stir the juice and the sugar together well, then pour into a fermentation jar and fit a bung and air-lock. The fermentation jar should be dark, since a transparent vessel will cause the colour of the wine to fade (a rule that applies to all red wines). Do not fill the jar right up to the top. Blackberry wine bubbles and foams as it ferments, so if you fill the jar too full the wine will overflow. Pour any excess liquid into a bottle and plug the neck with cotton wool – then, when fermentation has ceased (when there is no more bubbling), you can top up the jar from the bottle. Leave the brew in the jar until a sediment has formed at the bottom, with clear wine above it. Then siphon off the clear wine into a second fermentation jar and repeat the process, which is called 'racking'. Next time the wine clears it is ready for bottling. Nine months after you have bottled it, it will be ready for drinking.

Dewberry *Rubus caesius*

Very similar to the blackberry is the dewberry. The two are not easy to distinguish from a distance, but the dewberry is a smaller plant and the fruit ripens earlier. If you look closely, you will see that dewberries are made up of fewer but larger drupes, or segments, and that the sepals are always turned up round the base of the fruit (whereas blackberry sepals are usually turned down, away from the berry). Also, dewberry generally has fewer berries on each stem and the leaves are always composed of three leaflets with relatively finely serrated edges, while blackberry leaves vary considerably, having between three and seven leaflets, and the edges are usually broader-toothed.

Flowers of the Hedgerow

One of few advantages of the economic recession of the 1970s was the lack of finance to spray the hedgerows with herbicides, which allowed some of them to return to their former glory. For my great-grandmother the hedgerow, which she loved for its riot of summer colour, served as larder and medicine chest. Dandelion and burdock, self-heal and comfrey, foxglove and feverfew, colts-foot, teasel, daisy, nettle and yarrow, she gathered in season and put to good use.

Dandelion *Taraxacum officinale*

In mediaeval Latin the name for dandelion was *dens leonis* ('lion's tooth') and this became *dent-de-lion* in French. The leaf does indeed have a tooth-like shape and the tap-root resembles a canine tooth. The scientific name *Taraxacum* probably derives from two Greek words: *taraxos*, which means disorder, and *akos*, which means remedy. The medicinal virtues of the dandelion have long been appreciated

and it is without doubt an efficient diuretic. Many of its vernacular names reflect this characteristic, including pee-abed, pissimire, piss-a-bed and mess-a-bed.

Dandelion wine has a good flavour and steel workers at one time drank gallons of dandelion beer in an effort to replace the moisture lost through perspiration thanks to the heat of the furnaces. The leaves were and are used in salads, and dandelion and burdock is still one of the most popular mineral water flavours.

Dandelion seed heads are usually called 'dandelion clocks' and are puffed at by children everywhere, chanting 'One o'clock . . . two o'clock . . . three o'clock . . . etc., until all the seeds have been blown away and the time calculated. For this reason the plant is sometimes called 'one, two, three' and 'child's clock'. When all the seeds have been blown off you are left with a 'bald pate', which accounts for yet another vernacular name, 'monk's head'.

The plant itself is a member of the Compositae (daisy) family. Dandelions have a thick tap-root, from which arises a rosette of leaves without stems. The flower heads rise on hollow stalks from this rosette. Sadly, the dandelion is too common for its beauty to be appreciated. Once picked and brought indoors out of direct sunlight the flower closes up and its splendour is lost. The flower heads consist of many small structures called florets and if examined closely each of these can be seen to be a perfect flower in its own right, possessing both anthers and stigmas. At the end of the flowering period the seed develops a parachute which supports it and enables it to be transported considerable distances. The leaves contain a bitter-tasting compound called taraxacin which functions as a laxative. They also contain vitamins A

Dandelion

and C and in Russia the milky latex is extracted from the stems and good-quality rubber is produced from it.

Burdock *Arctium lappa*

When we were children, great battles raged during the shortening days of autumn as eager antagonists pelted each other with burs – a sport no less popular with children of other times, for Culpepper wrote that the little boys of his day loved to 'pull off burs to throw and stick upon one another'. In fact, the sharp hooks of the burs are designed to catch in the fur or fleece of passing animals, then are knocked off when the animal grooms itself. In this way they are often transported a considerable distance, so widespread distribution of the seed is ensured.

Gerard mentions burdock as an aphrodisiac, claiming that the young stems, peeled and either eaten raw or added to a stew, were guaranteed to 'increase the seed and to stir up lust'. Two of its vernacular names 'bachelor's buttons' and 'cuckold buttons' may refer to these aphrodisiac powers. Like dandelion, burdock was employed as a diuretic and doubtless that is why the two came to be so closely associated.

Burdock is another member of the Compositae family, as an examination of the flowers reveals. The leaves show a superficial resemblance to dock, hence the last syllable of its name. In Anglo-Saxon times it was called 'heriff', which comes from two words: *hoeg*, a hedge, and *reafe*, a robber. The resulting name, 'hedge-robber', picturesquely describes the tendency of the sharp burs to retain light material blown on to them by the wind. The *Arctium* part of the scientific name comes from the Greek word for bear, clearly referring to the rough texture of the burs. It is certainly a substantial plant, often reaching a height of 2 metres.

Burdock

Self-Heal *Prunella vulgaris*

The old botanical name for this plant, *Brunella*, was in use until about 1750, when *Prunella* became the accepted spelling. This may sound more pleasing to the ear but we have lost the true meaning, since *Brunella* derives from *Braüne*, the German for quinsy. It was so named, according to Parkinson, 'because it cureth that disease which they call die bruen common to soldiers in campe, but especially in garrison, which is an inflammation of the mouth, throat and tongue.'

The flower resembles the human mouth, tongue and throat, and self-heal was therefore used as a signature plant. Other names of significance to the old herbalists were 'carpenter's herb', 'hook heal', 'all heal' and 'sickle-wort'. At one time its reputation stood so high that according to Culpepper 'he needeth neither physician nor surgeon that hath self-heal and sanicle to help himself'.

Self-heal thrives on poor soil and is in

bloom from June throughout the summer. It is usually quite a small plant, but in ideal conditions can reach 30 cm in height. The stems, which are smooth and square, creep along the ground before rising upwards and the purple flowers grow in whorls of about six blooms. The shape of the flowers clearly shows the species to be a labiate and the corolla, made up of petals joined together, is composed of two lips. The upper lip has one round lobe and the lower lip three. The fruit consists of four little nuts which lie protected by the calyx until they are ripe. The plant does not entrust its future entirely to seeds, however, and its creeping stems can put down roots from any joint. Occasionally it also produces flowers, which never open but fertilise themselves – a process called cleistogamy, also employed by wood sorrel. Self-heal, although European in origin, has found its way to the United States, where it is known as 'heart of the earth' and sometimes as 'blue curls'.

Comfrey

Comfrey *Symphytum* spp

> The roots are so glutinative that they will solder or glue together flesh that is chapt to pieces. The same brushed and applied in the manner of a plaster doth heal all flesh and green wounds.

So wrote Pliny. Before him, Greek physicians had had a high regard for comfrey and used the roots as a 'vulnerary', or cure for battle wounds, and for internal injuries. However, Culpepper noted that 'a decoction of the leaves' is 'not as effectual as the roots'. Believed to be a native of Southern Russia, comfrey was probably brought to Europe by returning Crusaders and its English names of 'healing herb' and 'knitbone' have now entered the vocabularies of North America, Australia and New Zealand.

Recent medical research has revealed the presence of a compound containing nitrogen called allantoin, which speeds up cell growth and is an ideal substance for healing wounds. The little witch used to boil the leaves until they looked like spinach – indeed she ate comfrey chopped with butter – then bound them to a bruise with a leather belt. The leaves set quite hard and 'knitbone' was used to encase broken bones in the days before plaster of Paris. Bathing the feet in comfrey tea was said to relieve gout and, taken internally, it was used as a cure for stomach and intestinal ulcers. The value of the plant cannot be doubted, since in addition to allantoin it also contains useful glycosides, inulin, tannin and a great deal of soothing mucilage, plus a pleasant-tasting oil.

Coltsfoot *Tussilago farfara*

This delightful member of the Compositae family puts out its yellow flowers early in the year, often as early as January, before the appearance of the leaves, the shape of which accounts for its vernacular name. It was the roots, however, that the little witch and I used to collect. After washing off the loose earth and rubbing away the fibrous coat with sandpaper, we boiled the roots in honey and made delicious-tasting coltsfoot rock, which is still the most pleasant way I know to cure a sore throat.

I always thought that Sir Walter Raleigh brought back the habit of smoking to Europe, but Pliny recommends inhaling the fumes from coltsfoot leaves through a pipe made from a stout reed as a sure cure for a cough. Before the days of matches the cottony seeds and the white hairs scraped from the undersides of the leaves were soaked in saltpetre and used in tinder-boxes.

Foxglove *Digitalis purpurea*

The special indications which are commonly accepted for the administration of digitalis are increased pulse rate – say over 80 per minute – weakness of the contractile force of the heart muscle, cardiac dilation, and anasaroa, the cardinal indication being the failure of tonicity. The drug is more generally useful in mitral than in aortic cases, especially with scanty, high coloured urine, and with dropsy. . . . The drug

Coltsfoot

Foxglove

is often useful when mitral regurgitation has become super added to aortic disease and also in many cases of fibroid degeneration of the myocardium, but in the latter group of cases the effects should be carefully watched.

This extract is taken from *Diseases of the Heart* by Frederick W. Price, a medical textbook published in 1918. It underlines the golden rule of herbal medicine – that the correct dose is essential, or the patient will be killed, not cured. The drug digitalis is still useful in the treatment of heart diseases, but must be administered by a skilled doctor. The foxglove, with its distinctive purple bell-like flowers spotted with lighter colours, can grow to as much as 1½ metres in height and is a biennial. It is common in Britain and Europe, whilst in North America it has spread from cultivation. In addition to digitalis, all the foxglove's organs contain other dangerous cardiac glycosides which cause pains in the chest and abdomen, nausea and diarrhoea. Digitalis slows down the beat of a normal heart and increases the strength of the beat to such an extent that it can prove fatal. The little witch knew when to leave well alone and avoided all contact with 'dead fingers' or 'Devil's bells', as she called foxgloves.

Teasels *Dispsacus* spp

The small teasel (*Dispsacus pilosus*) is the species most frequently found in hedgerows, although fuller's teasel (*Dispsacus fullonum*) often escapes from areas where it is cultivated. 'Teasel' is derived from the Old English for 'to pluck', as the dried flower heads were once used for 'teasing' wool in order to separate the fibres before spinning. They were also used to raise the nap on cloth, a function now taken over by steel brushes, and are still used in the manufacture of billiard table coverings, the nap on the baize being raised with the hooked spikes.

The plant was known to and used by the Romans, who called it 'the basin of Venus,' since the leaves, which fit close into the stem and trap water, look like a washing bowl.

Fuller's teasel can reach a height of 2 metres. The long, pointed leaves make food by photosynthesis in the normal way, but recent research suggests that the plant may be partly insectivorous, since it absorbs nitrates from the decomposed bodies of flies that drown in the water of the 'basins', their bodies being dissolved by an enzyme produced by the leaves.

Teasels are usually biennial and the flower heads can be cut in August. My great-grandmother knew one area where the small teasel grew. She used the heads to bring up the nap on the black velvet hat which she kept specially for funerals. I still use them to good effect on suede shoes.

Daisy *Bellis perennis*

Daisies were once used to treat wounds, probably because, like self-heal, they soon flower again after cutting, quickly 'healing' their own 'wounds'. The juices from the flowers and leaves were used to treat ulcers in the mouth and sore or runny eyes, and a brew made from the roots was thought to cure fevers.

The name of the plant was originally 'day's eye', because in the daytime, when it is open, the flower looks like an eye. Hence its use as an eye salve. *Bellis* comes from the Roman legend about a nymph called Belides, whose duty it was to look after the meadows. One day, while she was dancing with her boy-friend, she was spotted by Vertumnus – the god of the changing seasons – and when he tried to pursue her, she escaped by changing into a daisy. The idea behind this myth would

Clover

was a great deal more practical and used red clover (*Trifolium pratense*) to treat coughs and skin rashes. Her cough syrup was made by dropping the red blossoms into a boiling, bubbling mixture of honey and water. This was taken every couple of hours and brought quick relief. Her skin lotion was made by infusing both flowers and leaves. After a few minutes' simmering, once the lotion had cooled, it was dabbed on the uncomfortable places.

White clover (*Trifolium repens*), also known as Dutch clover, played an important part in the rotation of crops, which did so much to revolutionise agriculture. No soil, however rich, can remain fertile if crops are continually removed without returning nutrients to it. To combat this a crop of white clover was grown every four years or so, then ploughed back into the ground. Clover, like all members of the Leguminosae family, has root nodules in which live bacteria able to convert atmospheric nitrogen directly into nitrates, thus enriching the soil. As a result, white clover, although not significant in its own right, provides good growing conditions for many important plants, including wheat, potatoes, carrots, turnips and other vegetables.

seem to be that, since the daisy flowers throughout the year, it is blind to the changing seasons.

Clover *Trifolium* spp

Any plant with trifoliate leaves was revered in the Middle Ages as a symbol of the Trinity and therefore as a powerful protection against the forces of the Devil:

Trefoil, St John's wort, vervain and dill
Hinder witches of their will.

A four-leaved clover was especially lucky, as the extra leaf was supposed to impart the gift of second sight. The little witch

Yarrow *Achillea millefolium*

Yarrow was said to keep away witches. The name derives from the Anglo-Saxon word *gearwe*, which in Middle English became 'ye arrow', since the feathery lower leaves look like flights of an arrow. The plant is a member of the Compositae family and the description *millefolium* ('thousand-leafed') refers to the intricate structure of the leaves.

An old Suffolk rhyme runs:

Green arrow, green arrow, you bears a
    white blow.

Yarrow

If my love loves me, my nose will bleed
now.

The 'green arrow' was thought to cause
the nose to bleed, thus relieving pressure
and so easing headaches. In fact, the leaves
and flowers contain an oily substance rich
in achilleine, tannin, resin, chamzulene
and cineole, which are effective against
headaches, indigestion and stomach disorders, and if added to a steam bath help
clear up a cold. The leaves can be eaten
and make a tasty addition to a salad.

Charlock *Sinapis arvensis*
This common plant of the fields was one
of my great-grandmother's favourite

herbs. She always called it zenry, and the
word rolled off her tongue with a delightful West Country trill.

Zenry is also known as wild mustard
and the little witch eagerly searched for
its yellow blooms from April onwards. Its
four petals arranged in the form of a cross
indicate that it is a member of the
Cruciferae family. The little witch gathered the leaves, choosing young ones,
which are less bitter, and either chopped
them up and added them to salads or
rolled them up, boiled them and served
them like broccoli. She also prepared the
flower buds in the same way and they
were extremely tasty. However, care
should be taken not to eat the seeds, since
they can cause vomiting.

Charlock

8 Plants of the Hills and Moors

A walk in the Highlands with a spring gale still laden with icy droplets of sleet or hail demonstrates vividly just how tough our upland plants have to be. The heavy rainfall in such mountainous regions makes them catchment areas for rivers and reservoirs. As the water soaks through the soil, most of the chemicals are leached out, producing a poor habitat for plants. Nevertheless, the hills and moors are almost always clothed with vegetation, ranging from conifers and native hardwoods to small bushes, and the most exposed areas can boast several species of herbs, mosses and liverworts.

Native Hardwoods

It is surprising how trees such as ash, birch, hawthorn, whitebeam and even the occasional oak manage to survive in some upland areas. Of these, the mountain ash and to a lesser extent the whitebeam seem to be particularly good at surviving at altitude.

Rowan (Mountain Ash) *Sorbus aucuparia*
Mountain ash is so called because of the superficial resemblance of the leaves to those of common ash. However, as the name *Sorbus* indicates, the mountain ash is a member of the apple family. The word *aucuparia* comes from the Latin *auceps*, a fowler or birdcatcher. Crushed berries were once mixed with the sap from birch trees or with mistletoe berries to produce bird lime, which was used to catch the small birds that were such an important part of the diet of our ancestors. The word rowan derives from the Norse word *rogn*. This has the same root as rune, which means a charm, and the old verses may well have been named after the wood on which they were inscribed.

The white blossom appears in May and June, and the most delightful feature of the tree is the blaze of scarlet berries which adorn the branches in September and October. J. N. Lightfoot, writing in the middle of the nineteenth century, confirmed the connection between the rowan and the old green religions:

It may to this day be observed to grow more frequently than any other tree in the neighbourhood of those druidical circles of stones so often seen in the north of Britain; and the superstitious still continue to retain a great veneration from it, which was undoubtedly handed down to them from early antiquity. They believe that any small part of this tree, carried about them, will prove a sovereign charm against all the dire effects of enchantment and witchcraft. The cattle also, as well as themselves are supposed to be preserved by it from evil; for the dairymaid will not forget to drive them from the shealings, or summer pastures, with a rod of the rowan tree, which she carefully lays up over the door of the sheal bothy or summer house, and drives them home again with the same. In Strathspey, they make, on the 1st of May, a hoop with the wood of this tree, and in the evening and morning

Rowan or mountain ash

cause the sheep and lambs to pass through it.

Although the berries are rather bitter, they are too numerous and succulent not to have been used as food when times got hard, but usually only after the more pleasant-tasting fruits had been eaten. The berries were dried and ground to produce a second-rate flour, which was baked to produce bread. The juice was used to 'cure the spleen' and prevent scurvy, as it is rich in vitamin C. John Evelyn, writing in the 1660s, reported that in Wales a good 'ale' was made from the berries. Since the rowan belongs to the apple family, rowan wine was probably not unlike a poor-man's cider. The Scots have always preferred to distil rather than ferment and in the Highlands rowan spirit, a powerful firewater, was drunk until the early years of this century.

The only use the little witch found for rowan was to crush the berries with a little honeyed water and a dash of vinegar. This she served with fowl and it tasted very much like cranberry sauce, which is also made from an upland plant. I was recently talking to a retired gamekeeper who had worked all his life on a grouse moor, as had his father before him, and he told me that they used rowan jelly as a flavouring for the 'odd brace of grouse wot came our way'.

Mountain ash timber is fine-grained, easily polished and extremely tough, so it is ideal for turnery work despite the small size of the tree. The bark has a high tannic acid content. Rowan is often found growing in upland areas with yew. Both kinds of timber were used in the production of bows and in the reign of Henry VIII a law was passed to try to preserve the tree.

Whitebeam *Sorbus aria*
Whitebeam only copes with altitude if there are sufficient outcroppings of limestone, as it does not do well on acid soils. Like rowan, however, whitebeam seems to tolerate a smoky atmosphere, which makes it a popular choice as a town tree despite its more normal upland habitat. The tree is quite small and it is not profitable to grow it for timber on a large scale, but the wood is very strong and is ideal for cart axles and the cog wheels of flour mills. The name whitebeam comes from the colour of the undersides of the leaves. *Sorbus* indicates its membership of the apple family, while *aria* records the mistaken belief that the species originated in Aria, in Asia.

The clusters of white flowers are larger than those of rowan. The berries are green in August and change to orange, then red, in September and October. The berries, like those of rowan, were made into a type of bread during periods of crop failure.

Native Conifers

Britain is not too well off for native conifers, but the species that are indigenous have played a prominent part in folklore, commerce and medicine. The Scots pine, yew and juniper have all played a significant role in our social history, whilst in recent years the growing of alien conifers has become a major industry and is likely to become even more important in the future.

Scots Pine *Pinus sylvestris*
The pine woods of Scotland are still among the finest sights in Britain, but at one time other areas of the country too were densely clothed with 'fir trees', as they were once called. Rich in resin, the timber burns well and it may be that the

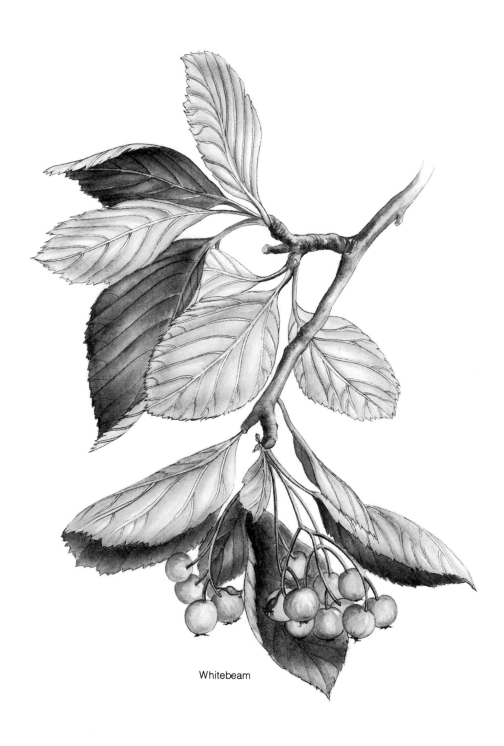

Whitebeam

words 'fir' and 'fire' had a common origin. Turpentine and tar are extracted from the Scots pine and related species, particularly the Carolina pine. A piece of bark is stripped from the trunk in spring and a hollow cut to catch the liquid. The lighter liquid obtained immediately is pure turpentine, while the thick residue is mixed with water and distilled to produce spirits of turpentine. The solid material left used to be sold until the beginning of the 1930s as yellow resin.

The Revd C. A. Johns describes how low tar was extracted from Scots pine:

The situation most favourable to the process is in a forest near to a marsh or bog, because the roots of the Scots pine from which tar is principally extracted are always most productive in such places. A conical cavity is made in the ground (generally in the side of a bank or sloping hill); and the roots together with logs and billets of the wood, being neatly trussed in a stack of the same conical shape, are let into this cavity. The whole is then covered with turf, to prevent the volatile parts from being dissipated, which by means of a heavy wooden mallet and a wooden stamper, worked separately by two men, is beaten down and rendered as firm as possible above the wood. The stack of billets is then kindled, and a slow combustion of the kiln takes place, as in making charcoal. During this combustion the tar exudes; and a cast iron pan being fixed at the bottom of the funnel, with a spout that projects through the side of the bank, barrels are placed beneath this spout to collect the fluid as it comes away. As fast as these barrels are filled, they are bunged and are then ready for immediate exportation. From this description it will be evident that the mode of obtaining

the turpentine, melted by fire, mixing with the sap and juices of the pine, while the wood itself, becoming charred is converted into charcoal.

Tar was produced by this method in ancient Greece and subsequently in other European countries, including Scotland, where usable conifers were plentiful. Stockholm tar, as it was called, was used for sealing the seams between planks on sailing ships and for weatherproofing roofs and wooden buildings. The traditional method of producing tar described above consumed a great deal of wood and modern methods pioneered in Scandinavia involving carefully controlled distillation in retorts are much less wasteful of timber.

Pine timber has been put to a variety of uses, including building, for which it must be well creosoted. Items as varied as organ pipes, arrows, roof timbering and baskets were once made from it. In recent years the production of telegraph poles has been one of its main uses. Packing cases, ladders, furniture and waggons have all been made from pine wood – and there have even been times of hardship when the poor have been forced to eat it, the inner bark being ground up and made into a rough form of bread.

Yew *Taxus baccata*

To my mind this majestic tree scarcely deserves its sombre reputation. It never grows very tall, seldom exceeding 15 metres in height, but its girth is remarkable and often exceeds 10 metres in diameter. The timber is of good quality, but because the trunk is broken up into columns called flutes it is not easily cut into planks. Nevertheless the heartwood, which is of a rich reddish brown, marked by the darker rings of summer wood, is still in demand by skilled cabinet makers.

Scots pine

This heartwood, but not the paler sap wood, is remarkably durable and yew fenceposts are reputed to outlast even those of iron.

The evergreen leaves can be eaten by cattle when fresh, but are poisonous once they have dried. The male and female flowers are usually produced on separate trees, though the occasional hermaphrodite plant occasionally occurs. The fruits look rather like tiny acorns, and are contained in red fleshy cups called arils. These should never be eaten since the seed protected by the flesh is extremely poisonous to humans. Birds, however, especially those of the thrush family, eat the arils and seem able to pass the seeds through their gut without ill-effect. Indeed, some authorities suggest that yew seeds which have passed through the gut of a bird germinate the following year, whilst those not so treated are obliged to wait an extra year. The slow growing yew often becomes hollow and a seed falling into the centre may germinate and grow inside the parent.

There are many possible explanations of the origins of both the scientific and vernacular names. Taxus may derive from *toxon* (a bow) or perhaps from *taxis* (a comb) since the leaves are arranged on the twigs like the teeth of a comb. It may even be that it derives from *toxicum*, meaning poison. Indeed, a dangerous alkaloid called taxine has been extracted from the leaves.

The Old English *iw*, meaning greenery, probably refers to the evergreen nature of the tree, which must have provided welcome shelter from the wintry weather. This may well be why the green religions held their meetings under yews. Later the Christians often built their churches close by, using the yews for shelter while the building was being constructed. This is surely a more plausible explanation of the fact that yews are found in churchyards than the one usually given. It seems unlikely that yews were deliberately planted in churchyards to provide staves for bows in case of emergency. By the time the stave was cut and fashioned into a bow the battle would have been lost!

Juniper *Juniperus communis*

Although juniper is less common in the wild than it used to be, it still thrives on some of the limestone uplands and also on the chalklands of Britain. The timber burns exceptionally well and, as the maximum height of the shrub is about 15 metres, it is felled easily. At one time juniper charcoal was called 'savin'. It made excellent gunpowder, which was much more predictable than other kinds and was therefore safer to use. It was this quality that caused so much juniper to be felled. The black berries, which are small but juicy, were once collected for flavouring gin. The Old French for juniper was *genevre*, from which, via the Dutch

Juniper

genever, we get the name 'gin' and at one time there was a profitable industry in Highland Scotland exporting berries to Holland. In Sweden the berries were eaten as a conserve or roasted and used to produce a drink not unlike coffee. The leaves, berries and timber have a delicate perfume, especially when burned, and were once used to fumigate the bedrooms of the sick. Juniper timber will take a high polish, especially when rubbed with beech-nut oil, and I still enjoy carving juniper wood to make thin, beautifully scented bookmarks.

Introduced Conifers

Norway Spruce *Picea abies*
When Britain and the mainland of Europe were still joined, before the melting ice raised the sea level and isolated our island, spruce may well have been present here. The timber is so versatile and yet so soft to cut that it was soon used up. Some experts believe that our climate was not suitable for the survival of the native variety and that Norway spruce was imported into Britain from the time of the Vikings. It was used for oars, masts and poles and in the production of what was known as Burgundy pitch. The resin was collected in the spring before the sap began to rise. An incision was made in the bark and the liquid collected in a cup and allowed to dry. It could be transported in this condition, then mixed with hot water to produce the pitch as and when required. There was some early confusion regarding the Norway spruce. Gerard, writing in 1597, and Evelyn, writing in 1662, both thought that it was the female of the silver fir. The timber of Norway spruce is known as whitewood and its bark was once eagerly sought by tanners. Up to the middle of the nineteenth century the leaves were used to add flavour to beer, especially in Germany. The word spruce derives from *prusse* meaning Prussian.

There will always be a demand for the timber. Indeed, no violin sounds right unless it has a belly fashioned from spruce and the soundboard of a piano is also of finely carved spruce. But perhaps the greatest demand these days is for Christmas trees, now a considerable business. Before Queen Victoria married Prince Albert the traditional British Christmas tree was the holly, but Albert preferred the tree brought from his native Prussia. Ever since, spruce has been regarded as the traditional tree.

The Norway spruce has soft needles that are green on both the upper and lower surfaces. The more recently imported Sitka spruce (*Picea sitchensis*), brought from North America and named after a town in Alaska, has sharper needles, which are green on the upper surface but a slatey-blue below.

Larch *Larix* spp
The Greeks and Romans built ships of larch, made the supports of bridges from the trunks, and used the bark for tanning leather and for the manufacture of turpentine. It is still a popular tree both for timber and for its beauty. It has the advantage of being a deciduous conifer, the orange of its dying foliage in autumn being exceeded in beauty only by its delicate spring greenery. As a rough guide, the needle-like leaves of the spruce are single; pine leaves are in pairs; and larch foliage is found in lumps, grouped together. Thus we have 's' for 'spruce' and 'single'; 'p' for 'pair' and 'pine'; and 'l' for 'larch' and 'lumps'. An easy way to sort conifers into the correct family.

The European larch (*Larix decidua*) was a popular tree with early planters and

Larch

Parkinson in his *Paradisus*, published in 1629, mentions that it was a rarity but was being increasingly planted. In the eighteenth century the Earls of Atholl turned their estates near Dunkeld in Perthshire into an immense nursery and it was here, perhaps by accident, that the first commercially valuable hybrid conifer was produced. A cross between the European and the Japanese larch (*Larix leptolepsis*), the Dunkeld larch possesses the former's ability to cope with wind and rain and the latter's resistance to disease, a notable example of hybrid vigour. The Dunkeld larch produces a ready supply of timber and is still used in boatbuilding and for pit props.

Shrubs

The upland shrubs include broom and gorse, heather, bilberry and bog myrtle.

Broom *Cytisus scoparius*

There are many species and sub-species of broom, but the most valuable to the herbalist is common broom, found both on upland and lowland heaths. It is often defined as a deciduous unarmed shrub – in contrast to gorse, which has sharp prickles. Broom derives its name from the Old English *brom*, a thorny shrub, although the prickles are quite soft. It can grow up to about 2 metres tall and its blaze of yellow flowers on a hillside is often visible for miles. The little witch was a great believer in the value of the flowers for those 'having trouble with their waterworks'. Broom has been found to contain an alkaloid, spartein, which is effective as a diuretic. Its use is not to be recommended, however, since some of the sub-species contain traces of much more dangerous compounds, which can cause serious harm in the hands of an unqualified or careless herbalist. The

old lady, despite her apparently haphazard approach, was both knowledgeable and careful and always used very small doses. She also maintained that a few fresh flowers of broom sprinkled in the bath water could bring relief from rheumatic pains. Broom has been used in the treatment of lung infections and thyroid trouble too. Again, strict medical supervision is essential.

The old name for broom was *Planta genista*. The coat of arms of Henry II were emblazoned with a spray of broom pods, since his father Geoffrey of Anjou hailed from an area in France where the plant is common, and it was from the old name for broom that the Plantagenets took their name. Broom was one of the great plants of love and magic. It is said that witches gave amorous ladies a potion of broom to slip into their husband's wine. This was no aphrodisiac, but made the husband sleep soundly while the errant wife crept away into the arms of her love.

Broom

Broom was also recommended for virgins who wished to preserve their innocence:

Take the blossom of the broom,
The blossom it smells sweet,
And strew it at your true love's head
And likewise at his feet.

The name 'broom' relates to its use as a sweeping brush, since the more substantial branches made ideal broomsticks.

Gorse *Ulex europaeus*

'Vuzz gathering' has long been a custom in Cornwall and I spent many a day gathering gorse with the little witch. The prickly gorse bushes were cut and made into faggots, which when burned produced a fire hot enough for baking bread and, in the old days, for burning lime and for smelting metals on a small scale. I use gorse as kindling to this day and find the prickly, aromatic leaves ideal for getting a wood stove going. Gorse, the little witch told me, was also good cattle food, but to make the prickles more palatable the branches were squeezed through a cider mill or an old-fasioned mangle.

Gorse or furze, as it is called in many parts of the country, is seldom without the odd yellow flower to brighten its dull greenery. In Scotland it is said that:

When the whin gangs out of bloom
Will mean the end of Edinburgh town.

A more romantic and more widely known saying is that:

When the whin is out of bloom
Kissing's out of season.

Gorse was often included in bridal bouquets, but the bride was forbidden to bring it into the house, since that was said to be unlucky. In Cornwall blazing bundles of 'vuzz' were carried around a herd of cows on Midsummer Eve to keep away the evil spirits and preserve the herd for another year.

Heather

There are three main species of heather, all of which commonly grow in conditions which kill off less hardy plants. Highlanders had to use whatever materials were available and so heather was pressed into service for brushes, thatching, fuel and ropes. Its roots were used to fashion dagger handles and the stems, mixed with cow dung, made excellent wattle and daub walls.

Ling or common heather (*Calluna vulgaris*) has a place in homeopathy and the fresh branches contain an essence used in the treatment of insomnia and painful joints. My great-grandmother dried the leaves to produce heather tea, which besides easing her rheumatism worked wonders for upset stomachs and diarrhoea. The glycosides contained in the

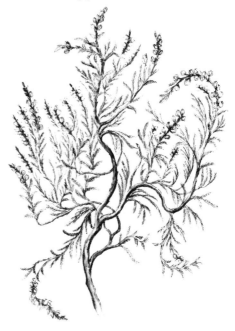

Common heather or ling

dry shoots and flowers also promote sweating and so bring down the temperature of flu victims. In addition, heather-scented baths are recommended for cystitis – or simply to revive weary limbs.

Ling is the most common of all the heathers and can be recognised by its tiny flowers, each composed of four petals covered by a long purple calyx. It is this which makes the purple heather moors such a glorious sight. The young shoots are a vital part of the diet of red grouse.

The other two common species of heather native to Britain are bell heather (*Erica cinerea*), which can be recognised by its bell-shaped flowers, and cross-leaved heath (*Erica tetralix*), which is identified by its spiky leaves arranged in fours, like a cross.

To my mind heather honey cannot be beaten for flavour. At one time ling was used to flavour ale and I always add a sprig or two to my home-brewed beer. I sometimes add a few sprigs to bilberry wine too, since the two plants grow so closely together.

Bilberry

Bilberry *Vaccinium myrtillus*

Bilberries are also known as whortleberries, huckleberries and whinberries and, in Scotland, as blaeberries. They are often found growing together with cowberries (*Vaccinium vitis-idaea*), which are called red whortleberries in many parts of the country. The bilberry is a deciduous shrub, but cowberry is evergreen. Both species have valuable medical properties.

The berries of bilberry are purple-black, as many a lass has discovered by sitting on a breezy moor in a new white dress. At one time the dark juice was used to make ink. Bilberries ripen during August and September, and those with the patience to gather the tiny fruits can drink the juice from the crushed berries, which are rich in vitamin C. Wine,

jellies, jams, pies and potions are all easily made from the berries or their juice and are a healthy addition to one's diet. Bilberries, which lower the blood-sugar level, are invaluable to diabetics and are said to cure diarrhoea and bedwetting, as well as stomach ache and nausea. The juice is also rich in carotene (pro-vitamin A) and is, therefore, good for your eyesight.

The berries are not the only part of the bilberry that can be used. The leaves infused in hot water are recommended for eczema of the hands and can also be used as a gargle to soothe an ulcerated mouth or infected throat. The roots too had a use and, chopped and shaken with water, served as an astringent to disinfect wounds and stimulate healing. The little witch was never without a supply of dried roots, leaves and berries, all of which she mixed with boiling water to produce an inhalant to clear the sinuses or a blocked nose, adding cowberry leaves and bog myrtle to the mixture.

The red berries of the cowberry were used as a treatment for diarrhoea and the leaves for arthritis, but because it contains the mildly harmful chemical arbutin, cowberry is best applied externally and should only occasionally be drunk as an infusion.

Bog Myrtle *Myrica gale*

Driving over the Pennines on a windy autumn evening, I stopped at a pub called The Sweet Gale. As I locked my car, the sweet aromatic scent of bog myrtle seemed to fill the air and I wondered how long it had been since gale beer was brewed in that part of Yorkshire. In his herbal, Gerard, who came from Cheshire, mentions that gale caused a frothiness in the ale and put a head on the brew 'fit to make a man quickly drunk'. In the days before fenlands were drained, bog myrtle must have been much more common than it is today. My great-grandmother never failed to keep us supplied with the leaves, which kept moths – and fleas, though we never mentioned them – at bay. We still sometimes burn a bunch of leaves on our log fire, which adds a marvellous aroma to the room. The fragrance is at its strongest during the flowering period, which lasts from April until the middle of July. In some areas bog myrtle was called sweet willow and a yellow dye was extracted from it for use in the tanning industry.

Herbs

The little witch always came back from the moors with her flower basket overflowing with herbs. Many a fascinating evening was spent at her feet listening to the flickering flame of her oil lamp and to stories which proved her green cunning. Lousewort, eyebright, tormentil, butterwort, bog asphodel and plantain were all often to be found in her basket.

Lousewort *Pedicularis sylvatica*
Red Rattle *Pedicularis palustris*

Lousewort usually occurs with red rattle, which is smaller, less treelike and often called marsh lousewort. Even Gerard could not find much value in either species, though he made part of his living selling plants and potions, and somewhat sourly wrote, 'What temperature or vertue this herb is of, men have not as yet been carefull to know, seeing it is accounted unprofitable.'

Red rattle is so called because when the seeds are ripe and the stalk sways in the breeze they rattle inside the fruit head. The origin of the name lousewort is less certain. According to the old lady, the perfume of the flowers was thought to attract fleas and lice. Before the nineteenth century, she said, when standards of hygiene were poor, a sprig of lousewort was often hung around the neck. Once the parasites had congregated on it, the sprig was burned. A much more plausible theory is that lousewort grows well on poorly drained soil where sheep are likely to pick up lice and other parasites.

Lousewort is itself a semi-parasite, obtaining part of its food by pushing suckers into the roots of grasses. The same is true of eyebright, another upland plant, like lousewort often found in the boggy slacks of sand dunes.

Eyebright *Euphrasia officinalis*

This delightful little plant with a clear yellow centre was bound to attract the attention of those who thought plants resembling human organs contained compounds destined to treat them. All the old herbalists were convinced of its value, as was my great-grandmother, and in this instance at least their faith was fully justified. Culpepper, never one to mince words, wrote 'If eyebright was as much

Eyebright

diarrhoea and also for 'the bloody flux', by which he probably meant the problems faced by women after childbirth and during menstruation. The treatment, he assures us, is even more effective if the powder is mixed with the water in which a blacksmith has quenched his iron. If the leaves and flowers are added to bathwater or are applied directly in the form of a lotion, tormentil eases sunburn. Added to toothpaste, it helps cleanse the mouth and get rid of mouth ulcers.

At one time a red dye was made from tomentil roots by the Scots of the Highlands and Islands. They used the dye as a substitute for oak bark in tanning hides and treated their fishing nets with it. An old and wrinkled inhabitant of the Hebridean island of Canna told me that his father used to tan his nets with 'red root' and found they lasted ten times longer.

used as it is neglected, it would have spoiled the trade of the spectacle makers.'

The little witch made a wonderfully soothing eyewash by adding a cupful of chopped eyebright flowers and leaves, together with a couple of tablespoons of heather honey, to a pint of warm water. She dabbed the solution on sore eyes with small pieces of lint and I still use the same mixture to relieve conjunctivitis. Taken internally, eyebright has been used to treat jaundice, but I think we are on much more dangerous ground here than in using it externally as an eyewash.

Tormentil *Potentilla erecta*
Tormentil was one of the little witch's favourite herbs, although she never failed to warn against using large doses. She used to dig up the roots and apply fragments of them to cuts as an astringent. Boiled in milk the roots also have been used in the treatment of colic in both farm animals and humans. The name derives from tormina (the term used by doctors in the seventeenth century for griping of the bowels) which came from the Latin *torquere*, 'to twist'. Gerard recommended the dried and powdered roots for treating

Butterwort *Pinguicula vulgaris*
Butterwort is able to survive in surprisingly hostile habitats by trapping and digesting insects. Its leaves, which have a rubbery texture, are covered with sticky, sensitive hairs. When an insect lands on the leaf it is trapped by the hairs and its struggles trigger the leaf to fold over and produce what amounts to a temporary stomach, into which pour digestive juices. These same juices were at one time used to make milk curdle. This property is today still exploited in Lapland to produce globules of energy-rich reindeer fat which are eaten to keep out the cold winter winds. In Britain butterwort was once important in the manufacture of butter and cheese, the chopped leaves being stirred into the milk, then picked out after it had coagulated. I have made cheese by this method and found the product excellent. The word *pinguicula* comes from the Latin for fat. The leaves

Butterwort

are in the form of a rosette from which the purple flower arises on a long narrow stalk.

Butterwort also had its medical uses. For colds and coughs, particularly for whooping cough, some herbalists still recommend two or three teaspoonfuls of the herb, freshly gathered, chopped up and infused in a cupful of hot water for a couple of minutes, then drunk or used as a gargle. It may well be that the large quantities of mucilage in butterwort soothe the throat, like a linctus. The little witch always told the young ladies of the family to bathe their skin in the mixture to improve its smoothness, which seemed to work. Some writers believe that because butterwort was rubbed on cows' udders to protect the milk against witchcraft the name is magical rather than functional. There is no reason why it should not have been both.

Bog Asphodel *Narthecium ossifragum*

At one time this charming yellow-flowered plant was known as *Asphodelus lancastriae*, the Lancashire asphodel, and was used in that county as a blonde hair dye. The old name was appropriately 'maiden's hair', whilst in Shetland it was called 'yellow grass'. After the yellow flowers with their contrasting red anthers have faded, the stems change to a saffron colour and it is from these that the dye is extracted.

Plantains

Both the great plantain (*Plantago major*) and ribwort (*Plantago lanceolata*) may be found in upland areas. Although they are ubiquitous, it is those growing in upland areas that account for the name of waybread. At one time the river valleys were wet, swampy and full of mosquitoes transmitting the ague (malaria). The safest trade routes followed the hilltops or high ground and plantain seeds, easily spread by the shoes of travellers, germinated to produce the sharp-tasting leaves of waybread, which provided the weary with a refreshing chew.

After describing how the medicinal effects of plantain are governed by both Mars and Venus, Culpepper goes on to extol the virtues of the plant:

The juice of plantain clarified and drank for divers days together, either of itself or in other drink, prevaileth wonderfully against all torments or excoriation in the guts and bowels, helpeth the distillations of rheum from the head, and stayeth all manner of fluxes, even women's courses when they flow too abundantly. It is good to stay spitting of blood and other bleedings of the mouth, or the making of foul and bloody water, by reason of any ulcer in the veins or bladder, and also

Ribwort

lungs, or coughs that come of heat. The decoction or powder of the roots or seeds is much more binding for all these purposes than the leaves. Dioscorides saith, that the roots boiled in wine and taken helpeth the tertian ague. . . . The herb, but especially the seed, is held to be profitable against the dropsy, the falling sickness, the yellow jaundice and stoppings of the liver and veins.

Culpepper was obviously a great believer in plantain and recommends extracts of the herb not only to fill rotting teeth and soothe sore eyes and ears, but also to ease the pain of 'piles in the fundament, kill worms in the belly, ringworm, itches in the head, shingles and inward or outward running sores'. Recent chemical analysis has revealed that the plant contains vitamin C, tannins, salicic acid and a great deal of mucilage, so it seems likely that some of Culpepper's treatments were effective and brought relief to his patients.

the too free bleeding of wounds. It is held an especial remedy for those that are troubled with the phthisic, or consumption of the lungs, or ulcers of the

9 Plants of the Riverside

My great-grandmother was not pleased when at last she was forced to admit that she had 'run out of ballast' and no longer had the wind to climb the steep slopes to gather plants from the moors and hillsides. She was ninety-three when she finally gave in. After that she spent much of her time wandering the river banks, gazing longingly upstream and talking about the plants that she knew grew 'up there in the wind'. But the waterside plants soon fired her enthusiasm and it was not long before her spirits recovered.

The riverside is a place of delight for both herbalist and naturalist. The banks of the river are dominated and graced by alder and willow. Riverside woodlands abound with herbs such as butterbur, nettle, mugwort, dog's mercury, golden saxifrage, ground ivy and soapwort. Succulent green water-meadows are resplendent with the golden blooms of kingcup and cowslip, and milky carpets of meadowsweet interspersed with lady's smock and marsh woundwort. And at the water's edge, often overlooked but no less important, are the rushes, reeds and horsetails.

Alder *Alnus glutinosa*

The alder is one of the most underrated trees of our countryside. It grows to a height of about 12 metres and is found along streams, rivers and in areas where the soil is damp. Its roots reach out beneath the ground in search of running water, since it is a thirsty tree, and small rootlets, pink in colour, actually project into the water. If these are examined under a microscope large numbers of swellings or root nodules can be seen and with a very high magnification the nodules are revealed to contain phenomenal numbers of bacteria. These bacteria play a vital role in the life of the alder, since they are able to convert the nitrogen from the atmosphere directly into nitrates. When the bacteria die these nitrates become available to the tree, which can therefore thrive in conditions that would be impossible for many other species. The Revd C. A. Johns, writing towards the end of the nineteenth century, noted that:

> . . . the alder does not take a high rank among our picturesque trees, but we must recollect that it often flourishes where no other tree would live and thus ornaments a landscape which would otherwise be tame and naked. It retains its leaves too, until very late in the year; and gloomy though their tone may be we forget this defect when nearly all other trees are bare.

The uses found for the timber of the alder reflect its mode of growth. Timber formed in a perpetually wet habitat does not function efficiently in dry conditions that remove the water and leave the timber far too soft and light to last, but alder is ideal for bridge supports, water pipes and pumps which are continually submerged in water. Consequently, before the days of durable metals and plastics it was a major crop for those involved in the construction industry.

Alder

human spirits held within the tree by some supernatural power. The clog-block maker, however, had no such fears, being a practical and often hard-headed business man. His business sense made him something of a conservationist and he realised that it was in his own interests to make the maximum number of blocks from the minimum volume of timber. He would also plant seedlings in suitable spots to ensure that a supply of timber was always available both for himself and for his descendants. Once fashioned, the clog blocks were taken to the villages and towns and sold to the clog maker, who made footware to measure for each member of the community.

Another craftsman who relied upon the alder was the dyer, who before the days of the industrial chemist obtained a yellow dye from the green shoots, a lovely deep red from the bark, a green from the freshly cut catkins and a delicate pink from the wood chippings.

Willow *Salix* spp

In Britain we do not have any native palms with which to celebrate Palm Sunday, but the goat willow (*Salix caprea*), more generally known as pussy willow, which is in full bloom over the Easter period, provides a substitute.

During the nineteenth century, when the mills and factories of the North were working flat out to satisfy the needs of the civilised world, the workers shod themselves and their children with clogs, the soles of which were made from blocks of alder. The clog-block maker would wander the countryside looking for alder timber of a suitable age and, having found it, would negotiate with the landowner for the purchase of the tree. Once the deal was struck the craftsman set up his tent and cut down the tree. When first cut the timber is white, but soon it becomes blood red, before fading to its permanent colour of pink. Some folk refused to fell alder, thinking that the 'blood' running from the wound came from tortured

The willow family is extremely complicated and many species interbreed freely, which makes precise classification difficult. For example, the common sallow or grey willow (*Salix cinerea*) is considered by many to be a sub-species of the pussy willow, but I believe it is better treated as a separate species. It is a tree of variable height, from 1 metre up to as much as 10 metres. Although the main flowering period extends from late March to the middle of May, there is often a second flowering period during September.

In *Othello* Shakespeare put the

Cutting osiers for basketwork

following words into the mouth of the ill-fated Desdemona:

My mother had a maid called Barbara.
She was in love and he she loved proved
    mad
And did forsake her; she had a song of
    'willow'.

For centuries 'to wear the willow' has meant to grieve the loss of a sweetheart. The association probably originated from the passage in Psalm 137 which relates the story of the Jews held in captivity by the Babylonians. They refused to sing the songs of their homeland while in exile and hung their harps on the willow trees that grew by the rivers of Babylon as a sign of their grief.

The willow was of great practical value to our ancestors. The very name 'willow' comes from the Anglo-Saxon *velig*, which means 'supple'. The shoots or 'withies' especially of the osier (*Salix viminalis*) were, and still are in a diminishing number of villages, used in basket making and wickerwork. I remember going with my great-grandmother to watch the 'swill man', as she called him, cut his sticks, strip off the bark and then sit in his shed and produce a shopping basket in a remarkably short time.

Willow leaves are long and narrow, being as much as 10 cm in length. Many varieties have crinkled leaves and in some species – especially the white willow (*Salix alba*) and its sub-species *coerulea*, from which cricket bats are made – the undersides are covered with downy white or silver hairs. When a breeze makes the leaves turn and twist, they flash alternately green and white – a lovely sight on a fresh May morning. Willow trees carry catkins of both sexes and pollination is performed in the main by bees.

The little witch was herself no mean basket maker, but it was the good medicine of the tree that appealed to her most. She followed Gerard, who used leaves, bark and seed directly to staunch blood. According to Gerard, when added to wine willow eases the wind and prevents vomiting and the leaves were said to 'stop the lust' in both men and women. In addition, Gerard recommended mixing the burnt ashes of the leaves with vinegar to cure corns and warts or to get rid of superfluous flesh. For my own part, I have banished dandruff by washing my hair in a rinse made from the leaves or bark mixed with water, preferably with the addition of a little wine. It is now known that willow bark contains aspirin.

Butterbur *Petasites hybridus*

This most fascinating plant is a member of the Compositae (daisy) family. The flowers appear before the leaves, and the flowering period is generally from March to May. However, I have found butterbur growing on northern river banks as early as mid-January. Male and female flower heads are carried on separate plants. The male flowers can be recognised by the fact that their florets have a tube-shaped corolla with five teeth on it. Within this are five stamens, which produce pollen grains. The female plant, of course, has no stamens. Instead, it has a style made up of two lobes, the function of which is to catch the pollen carried to it by bees in search of the copius nectar produced by the flowers of both sexes. The aroma generated by the often extensive riverside colonies is not pleasant to the human nose, but the bees clearly find it attractive.

Two varieties occur, namely the common pink and the much rarer white butterbur, the latter almost entirely restricted to northern England. Butterbur is found throughout Europe and in much of northern and western Asia, and has even been introduced into North America. Among its vernacular names, of which there are many, we find 'wild rhubarb', 'bog rhubarb', 'snake's rhubarb' and 'umbrella leaves' – which

Butterbur

perfectly describes the huge leaves that dominate the river banks long after the flowers have died.

The word 'butterbur' itself is meaningful, since the leaves were used to wrap butter before the days of grease-proof wrapping paper. *Petasites* derives from the Greek *petasus*, a wide-brimmed hat – another reference to the rhubarb-like leaves. Butterbur promotes sweating and in the Middle Ages it was used against cholera, plague (including the Black Death) and other 'fevers'. In Germany it was known as pestilence-wort. The leaves were pulped, heated and used as a poultice in the treatment of boils and abscesses.

Stinging Nettle   *Urtica dioica*

Urtica is certainly an appropriate name for a plant with such a powerful sting. Nevertheless, the nettle was a useful plant in times gone by. Surprisingly, it has proved extremely difficult to cultivate, because it demands a soil exceptionally rich in nitrogen. It therefore grows well by river banks where rotting vegetation washed downstream has become trapped.

The leaves consist of a tough fibre which can be converted into cloth. In fact, an analysis of German army uniforms worn in the First World War revealed an 85 per cent presence of nettle fibre. In the past young nettles were eaten as a green vegetable and were often grown commercially for this purpose. The plant was also used to make nettle beer. Indeed, it was quite possible in the old days to sleep between nettle sheets, eat nettle as a vegetable from a nettle tablecloth, and assist the digestion with a glass of nettle beer.

Stinging nettles, like the elm tree, to which the nettle is closely related, have two methods of reproduction. There is a vegetative (or non-sexual) method, in-volving growth from underground rhizomes. When these rhizomes become fragmented the parts take up a separate existence. Nettles also reproduce sexually. The male plants produce clusters of flowers that stick out more stiffly, while on the female plants the flower clusters hang down.

The role of the nettle in the life of man has now changed from that of a vital plant difficult to cultivate to that of a weed, spurned and unwanted. But it is worth tolerating the nettle in our fertilised, nitrogen-rich parks and gardens, because it provides food for some twenty-seven species of insects, including many butter-flies, among them the small tortoiseshell and the peacock.

Mugwort   *Artemisia vulgaris*

In days gone by this aromatic plant was thought to be a cure-all. A perennial plant common throughout Britain, it flowers from late summer until well into the autumn. Mugwort may perhaps derive its name from the fact that it was used to give the bitter taste to beer before the cultivation of hops. Recognised through-out Europe as a powerful plant, in pre-Christian times it figured in a spell known as the 'Lay of the Nine Herbs' invoked as protection against witchcraft.

The merits of mugwort were still being expounded in fairly recent times. Accord-ing to *The Herbal Book* of 1867, which no doubt follows earlier herbals:

Mugwort removes obstructions of urine caused by stone. A decoction is said to cure the Ague. The Chinese use it to heal wounds applying the fresh plant bruised. A drachm of the leaves powdered was used to cure fits by Dr Home. Made into an ointment with lard and a few daisies it was good for boils. Three drachms of the powder boiled in wine is a speedy and certain remedy for sciatica. A decoction with chamomile and agrimony and the place bathed whilst it is warm takes away the pains of the sinews and cramp.

Dog's Mercury *Mercurialis perennis*
Local names for this plant are numerous, but among the more interesting are boggart flower, dog's flower, dog's medicine and snake's flower. 'Dog', as in dog rose, simply means 'common'. As its scientific name implies, dog's mercury is a perennial. An insignificant-looking plant, it grows freely in woods and is also found in the cool, shady, damp places that were thought to harbour poisonous snakes, goblins and boggarts.

An uncommon relative of dog's mercury, annual mercury (*Mercurialis annua*), was used by Gerard for making enemas and dog's mercury has an even more dramatic effect on the digestive system. The name mercury was given to the plants because the god Mercury was said to have discovered their medicinal properties.

If eaten, dog's mercury can be extremely dangerous and can cause severe irritation to the gut, in some cases leading to death. The mercuries are closely allied to the spurges, the two families containing related poisons. It is surprising how cattle, both wild and domesticated, know what is good for them. Cattle always leave dog's mercury alone, whilst seeking out other plants in the shady, succulent places where dog's mercury grows. Indeed, this is probably why poisons evolved in plants. After all, the more unpleasant the consequences of eating a plant are, the less likely it is to be consumed. Annual mercury is less poisonous than dog's mercury – which may explain why it is not so common – and in the not too distant past it was used as a

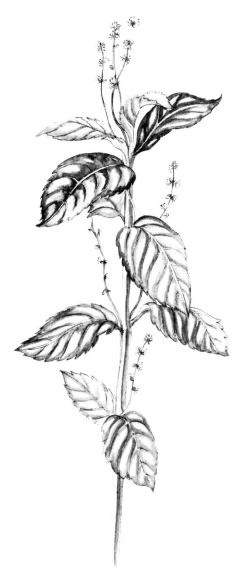

Dog's mercury

pot herb.

Male and female flowers are carried on separate plants, the females being smaller and much less common than their male partners. However, the shortage of female plants is not a great problem, since individual plants reproduce vegetatively from an extensive underground system of rhizomes.

The Greek writers Dioscorides and

Theophrastus both knew that the male and female flowers were carried on different plants and thought that the sex of an unborn baby could be determined by careful doses of male or female dog's mercury for three consecutive days after conception – with what consequences one dreads to think!

Golden Saxifrage

Chrysosplenium oppositifolium

Vernacular names for this tiny plant include ladies' cushion, buttered eggs, creeping Jenny and golden stonebreak. In parts of eastern France it was used in salads and was called *cresson-de-roche*, or rock cress.

In Britain the plant was prescribed as a remedy for melancholy and disorders of the spleen. Indeed, this is the origin of its scientific name *Chrysosplenium*, which derives from the Greek *Chrusos*, gold, and *splen*, spleen. Hence another of the plant's common names, golden-spleenwort. The word *oppositifolium* refers to the arrangement of pairs of leaves along the stem. This distinguishes the species from the rarer *Chrysosplenium alternifolium*, the leaves of which are arranged alternately.

Golden saxifrage is a very common plant in damp woods and along stream sides, but it is often overlooked because the tiny, attractive yellow flowers, which lack petals and are set against lush green leaves, are only about 4 cm across. Golden saxifrage is in flower from March to July, with the peak of the flowering period being in the middle of May.

Marsh Marigold *Caltha palustris*

This plant has over fifty vernacular names and was doubtless well known to our forebears because of its large golden blooms. In the days when much of Britain was covered by swamp and damp clouds of mist hung heavily over the land, the marsh marigold reflecting the light of the penetrating sun must have been a delight, as one of its vernacular names 'meadow-bright' suggests.

It has been discovered by examining pollen grains obtained from peat deposits that the plant was present in Britain before the ice ages and managed to survive in those bleak times in a few sheltered refuges. It still thrives in Iceland, so clearly it has retained its hardy streak. In Britain the full flush of growth comes during April and May. As a result, it often played a vital role in the fertility rites celebrated on May Day ('May flower' was one of its vernacular names). In Ireland, cowsheds were decorated with marsh marigolds in order to protect the animals from the evil influence of witches.

The species seems to have had few medicinal uses in Britain. Gerard wrote that 'touching the faculties of these plants we have nothing to saie, either out of other mens writings or our owne experience.' It was used, however, to produce a yellow dye and there are occasional references in Anglo-Saxon literature to it as a cure for skin rashes. From my experience it is more likely to create a rash than cure one. In the United States the unopened flower buds are used in spring salads and the vernacular name for the plant there is cowslip. Despite its lack of medicinal virtues, Gerard in his herbal of 1597 wrote this appreciation of the plant:

Marsh Marigold hath great broad leaves, somewhat round, smooth of a gallant greene colour slightly indented or purple about the edges, among which rise up thicke fat stalks, likewise greene; whereupon do growe goodly yellow flowers, glittering like gold.

The marsh marigold, in spite of its splendid colour, is totally without petals, and the function normally carried out by petals is performed by the large yellow sepals. The scientific names are aptly descriptive. *Caltha* comes from the Greek *kalathos*, a cup, while *palustris* is a Latin adjective meaning marshy. The vernacular names include 'water babies', 'water goggles', 'golden kingcup', 'soldier's buttons' and 'bog daisy'.

Ground Ivy *Glechoma hederacea*
Ground ivy was one of the little witch's favourite plants. I often wondered if she called it 'ale-hoof' because of the shape of its leaves, but it now seems fairly certain that the plant was used to flavour ale before hops. She not only added it to home brewed ale, but also made a rather bitter-tasting tea by infusing the dried leaves. I still find drinking this effective against a heavy cold or congested chest. Septic wounds are quickly cleaned by boiling any part of ground ivy in water to which honey has been added and washing the infected spot with the healing mixture. Ale-hoof was added to bathwater at least from the Middle Ages and was held to be very health-giving. As well as growing on the floor of riverside woodlands, this member of the catmint family, which is unrelated to true ivy, is also common on waste ground. It is a perennial and flowers early in the spring. The stem and leaves are hairy and the flowers bright purplish-blue, with a corolla tube almost 2½ cm long. A carpet

Ground ivy

of flowering ground ivy early in the year is often one of the first signs of spring.

Soapwort *Saponaria officinalis*

To washerwomen it must have seemed almost like manna from heaven when soapwort was discovered growing by the riverside. Rubbed with water, it produces a fine lather and by the time the Pilgrim Fathers had established themselves in America soapwort had already been transported across the Atlantic, where it became known as 'my lady's washing bowl'. In England one of its local names is 'Bouncing Bett'. The plant contains saponins, which have a much more gentle action than inorganic soap, and soapwort is still occasionally used by those restoring tapestries. Where large stands are found near the foundations of a ruined house, you may be sure you are near the site of the old laundry. A plant as useful as this pink-flowered perennial must also have been planted in cottage gardens, and it still grows along river banks.

Meadowsweet

Soapwort

Meadowsweet *Filipendula ulmaria*

The flowers of meadowsweet are rich in iron and magnesium, both of which are essential for the production of red blood cells. As a result, those who are anaemic, including women lacking in minerals following childbirth or menstruation, benefit from drinking a daily wine glass of the dried flowers infused with water. The recommended mixture is about an ounce (28 g) of flowers to a pint (½ litre) of water.

With its erect white blossoms ornamenting ponds, ditches and damp meadows and cascading from the hedges during the late summer, the plant well deserves its name of 'queen of the meadows'. The flower heads can stand over 1 metre high and because of their sweet scent were strewn on the floors of churches and houses before the days of flagstones or floorboards. The reason for their delightful aroma is not far to seek – for a close look at one of the hundreds of tiny flowers that make up each flower head reveals the typical five petals of the rose family, to which the plant belongs. The flowers have also been used for flavouring drinks, including mead, the honey-based brew which was such a favourite of our forefathers.

Meadowsweet was one of the three plants most favoured by the Druids and in view of its many uses it is difficult to understand the belief that to bring meadowsweet into the house brings death. The little witch dried the flowers in the garden shed, but nothing would persuade her to bring a living spray into the house in case it put us into a deathlike sleep.

Lady's Smock *Cardamine pratensis*
The leaves of this plant were once eaten to prevent scurvy, as they are very rich in vitamin C. They were also eaten in salads and the dried and powdered flowers were used during the eighteenth and nineteenth centuries to treat epilepsy and to get rid of vermin.

The origin of the name lady's smock has been the subject of much speculation, but it is possible that 'smock' is derived from Old English and may mean a 'lecherous glance'. The plant was probably used in the old May Day fertility ceremonies and 'May flower', another of its names, indicates this origin. So, probably does 'cuckoo flower', though the name may refer to the fact that the bird is first heard about this time. I discount the later suggestion that the flowers looked like a group of newly washed lady's

smocks laid out in the sun to dry. Certainly the connotation of lechery fits an anonymous Irish poem, dating from the fifteenth century:

Tender cress and cuckoo flower
And curly-haired fair-headed maids,
Sweet was the sound of their singing.

Another of the vernacular names, 'pick-folly', indicates that in some places it was considered unlucky to pick the plant at all. This was also the belief of the French, who avoided lady's smock because they thought that it was the favourite food of the adder and anyone who stole a bloom would be bitten and die before the year was out. In Germany those brave enough to pick it risked storm and pestilence being brought upon their dwelling place.

Lady's smock is in flower from May at least until August and a very pretty sight it is. The pale lilac or even white flowers carried on stalks up to 50 cm high, have four petals arranged in the form of a cross, since it is a member of the Cruciferae family. A common plant throughout Britain, especially where the ground is damp, it is often found at altitudes over 1,000 metres. Its favoured habitat, however, is along damp meadowlands, often accompanied by the blossoms of marsh woundwort and water mint.

Marsh Woundwort *Stachys palustris*
This species, as its name suggests, once played an important part in the treatment of wounds caused by accidents or in battle. The family is widespread and hedge woundwort (*Stachys sylvatica*) and field woundwort (*Stachys arvensis*) contain the same healing elements. There are about 200 species of woundworts occurring in most parts of the world except Australia and New Zealand. They may

Marsh woundwort

urine and helps all joint aches. It helps all cold griefs of the head, the vertigo, falling sickness, the lethargy, the wind colick, obstruction of the liver and spleen, stone in the kidneys and bladder. It provokes the terms, expels the dead birth: it is excellent good for the griefs of the sinews, itch, stone, and tooth-ache, the biting of mad dogs and venemous beasts and purgeth choler very gently.

Gerard knew the plant more as a wound herb, the leaves being applied directly to the wound, and claimed that it effected remarkable cures for those injured in brawls or duels.

Cowslip *Primula veris*

Cowslip is a polite way of saying 'cowslop' for it was thought, perhaps correctly, that it grew best where the cow had 'lifted its tail'.

Some amusing ideas surround the plant and it has a deep association with our folklore. It was thought that if the leaves were planted upside down the flowers would either turn out to be primroses or else would be coloured red, instead of yellow. Cowslip flowers were rolled into a ball and, in Wales especially, the young maidens of the village would form a circle. The ball was thrown from one to the other while they chanted:

Tisty, tasty tell me true,
Who shall I be married to?

The names of all the eligible bachelors were added to the rhyme. When the ball was dropped, a match was thought possible between the last-mentioned boy and the girl nearest to the cowslip ball.

From a medicinal point of view, cowslip was considered to have so many virtues that it was referred to as

either be annual or perennial, as in the case of marsh woundwort, and tend to grow from a basal rosette. When crushed, the broad heart-shaped leaves, which have a texture rather like flannel, emit an unpleasant, musty smell. The flowers are reddish-purple and are supported on a square stem.

Culpepper had such faith in marsh woundwort that he called it all-heal, recommending it because:

It kills the worms, helps the gout, cramps and convulsions; provokes

'St Peter's keys of heaven'. More significantly, it was known as palsy-wort since it was used in the cure of paralysis. It was also thought to cure migraine and amnesia, and the consumption of liberal quantities of cowslip wine was believed to prevent insomnia – as do elderberry wine, whisky, gin and best bitter! But the little witch, for one, was a great believer in cowslip wine.

Modern herbicides have now made the species rare in many areas and it is in need of the protection of the law.

Rushes, Reeds and Horsetails

Rushbearing

Rushbearing had its origin at the time when Christianity was introduced into England, Pope Gregory IV, in AD 827, directing that on the anniversary of the dedication of the Christian churches wrested from the Pagans, the converts to Christianity should "build themselves huts of the boughs of trees about their churches, and celebrate the solemnities with religious feastings." But when Christianity became firmly established, the annual rejoicing over the gain of the churches from the Pagans ceased to have that special significance. The festival, however, remained and served not only as a village festival but as a means of supplying rushes to be strewn on the church floor during the winter – hence it came to be known as "Rush-bearing".

In Anglo-Saxon times and far into the Middle Ages, the floors, even in royal palaces, were covered with rushes in place of carpets, and as the floors of the churches were either of stone or beaten earth, they were bitterly cold during the winter; the villagers there-fore, in strewing the floor with rushes, were providing for their own comfort. It was the custom for the young men of the village to gather rushes when they were at their full length, and piling them high up on carts, sometimes to the height of ten or twelve feet, take them to the church. The carts were gaily decorated with ribbons; the procession to the church was headed by music, and the day was given up entirely to merry making. In the course of time the decoration of the rush-carts became more and more elaborate.

This account written by Frank Hird in 1911 records the important ceremony of rushbearing in the now industrial town of Rochdale in Lancashire. Rushbearing was a feature of British country life until more elaborate and permanent floor coverings were devised. In some places – in Ambleside in the Lake District, for example – the tradition is still maintained today, even though rushes are no longer used on floors.

It was not only for floor coverings that the villagers relied on rushes – for once they had been peeled, the highly absorbent centre sections soaked in animal fat made excellent lights. The soft rush (*Juncus effusus*) was best for this purpose, but any species of marshland rush could be used and every villager gathered rushes during the autumn in preparation for the long, cold winter nights ahead. In his book *Cottage Economy* published in 1823 William Cobbett expresses his faith in the traditional rush light:

I was bred and brought up mostly by rush light, and I do not feel that I see less clearly than other people. Candles certainly were not much used by English labourers' dwellings in the days when they had meat dinners and

A thatcher at work

Sunday coats. Potatoes and taxed candles seem to have grown in fashion together.

As Cobbett hints, the rushes were soaked in the fat or dripping from the Sunday roast, the one 'meat dinner' of the week – so unlike candles, rush lights cost nothing.

Rushes provided not only a soft floor covering and a surprisingly bright light, but also, in the days when thatching was widespread, a dry roof over one's head. Thatch was only replaced by slate when canals, railways and finally road transport became efficient enough to carry roofing slates cheaply and without damage. Thatching and weaving were at one time two of the most important rural industries. Fortunately, both these skills are far from dead and are still being practised today.

Thatching and Weaving

Many plants can be used for thatching. Birch twigs, broom and heather have all been used in areas where they are common. In Anglesey marram grass was used. Ideally, thatching material should be light yet strong and easily obtainable. Wheat straw with its hollow stem is perfect. Rushes are light enough for thatching

Common reed (**top**), bulrush (**left**) and soft reed
(**bottom**)

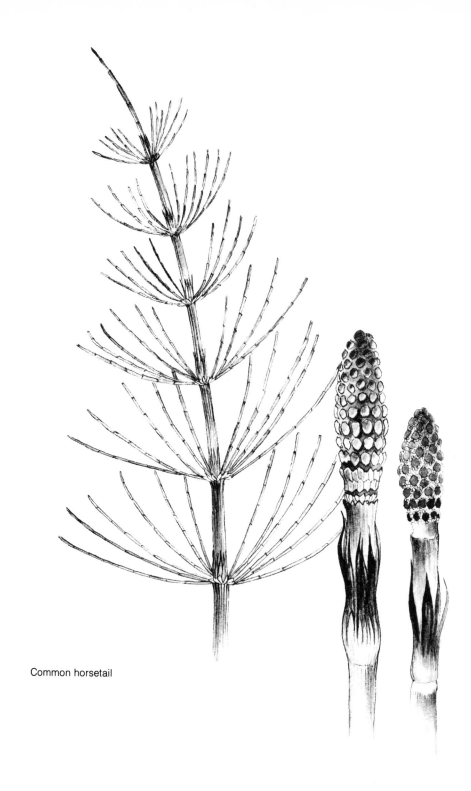

Common horsetail

too, as is the common reed (*Phragmites australis*), which grows in the marshy areas skirting rivers. The common reed flowers during August; by December the leaves have died and the dead stems are ready for harvesting. This is not wasteful because the underground rhizomes, which store food, are not affected by cutting and are ready to fuel next year's growth. The brittle stems break easily, but this does not worry the experienced thatcher, who piles the reeds on to the roof, then skilfully plaits them and clips them into place.

The common reed is, however, useless to the weaver, who requires a much more flexible material such as willow or the true bulrush (*Scirpus lacustris*) – not to be confused with great reedmace (*Typha latifolia*), which is also popularly known as bulrush. The supple bulrush stems can be cut from June until early September, then stacked till the weaver needs them. Before he starts work the weaver sprays the stems, which soak up the water and become pliable once more.

Horsetails *Equisetum* spp
The horsetails dominate the lower half

metre or so of the marshy areas of ponds and rivers. In the Carboniferous period, 250 million years ago, when the climate of Britain was much hotter and wetter than it is now, the horsetails grew to a height of 30 metres or more. When these 'fern trees' decayed and tumbled into the swamps, the weight of other fallen vegetation pressed them to the bottom and squeezed out the water, thus producing coal. Plants of the horsetail family contain appreciable amounts of silica. As a result, they can be used like sandpaper and watchmakers used to give a final polish to delicate working parts with them. At one time horsetails were known as pewter-wort since, before the days of steel wool and scouring powder, housewives and scullery maids used them for scrubbing pans and for polishing silver and pewter.

My great-grandmother never returned home from a picnic with dirty utensils. 'Pan scrubbers grows everywhere for them with the green cunning to foind 'em,' she said. Nor did she ever return from her wanderings empty-handed. Her green cunning kept her mind active and her body healthy. Both served the little witch well for over a hundred years.

Collecting and storing plants for herbal and culinary uses

The old herbalists believed – in many cases with good reason – that the valuable properties of wild plants were superior to those of their cultivated counterparts because nature herself had provided them with the right conditions and climate.

Plants growing at a distance from their neighbours are generally stronger than those crowded together, since they do not have to compete with one another, and according to Van Helmont, the seventeenth-century scholar, doctor and alchemist, plants should be gathered 'at the moment of their full maturity and greatest vigour' – at what he called 'the balmy time'. The time of day is important, too. A warm summer afternoon is an ideal time to be out gathering flowers and dawn and dusk collecting should be avoided, as dew laden plants quickly become mildewed.

Only a few leaves should be gathered from each plant or the plant will be killed and even common weeds should be picked sparingly. Some conservationists these days maintain that plants should not be picked at all, but who can object to a few blooms of dandelion, a handful of nettles or a bunch of ash keys? Timber or bark from wind-felled trees is likewise a harmless acquisition.

Once collected, plants should be kept in a dry, well-ventilated spot away from light and should be spread out flat in a single layer, never one on top of the other. If you have to put the plants in a container temporarily while you are out (a large wicker basket is ideal), remove them from it as soon as you return home. Try to collect only clean plants, since the vegetative parts should not be washed. With so much lead in the atmosphere, collecting near main roads is therefore to be avoided. During the drying process, turn the plants frequently. Then, when they are completely dry, cut them up. Roots require different treatment. After brushing off the soil, wash them, slice them into pieces, then leave them to dry. Once it has been dried and cut up, the plant material should be removed from direct contact with air and either stored wrapped in linen, in well-stoppered bottles or in airtight tins or plastic containers. Polythene bags are useless for this purpose.

It is important when using herbs to distinguish between infusions, decoctions and macerations. For all three you will usually find instructions in the herbal or recipe that you are following regarding the amount of material and quantity of liquid to use. An infusion is made, just like tea, by placing the dried material in a teapot, pouring on boiling water and leaving it to steep. To make a decoction, you simply place the material in cold water in a pan (not aluminium) and bring the mixture slowly to the boil, then simmer and strain. For a maceration, you leave the material to soak in liquid in a glass vessel. Water is not usually suitable, since macerations are often stored and water is not a good preservative. Oil, wine, vinegar or alcohol, on the other hand, are ideal. Once the material has had time to soak thoroughly, squeeze the mixture through a clean cloth to force out the juices from the plants. The type of wine used is crucial, since red wine is rich in tannin and has an astringent effect, whilst white wine lacks this property but is an efficient diuretic. The fluid should be stored in well-stoppered medicine or wine bottles, which have been rinsed and dried carefully, and should be kept away from strong light, preferably in a cool place.

Bibliography

ALLABY, M. *The Survival Handbook* (Macmillan, 1975)

ARNOLD, J. *The Shell Book of Country Crafts* (John Baker, 1968)

BARRACLOUGH, D. *A Flower-Lover's Miscellany* (Warne, 1961)

BEALS, K. M. *Flower Lore and Legend* (Holt, New York, 1917)

BLAIRE, J. L. (ed.) *British Wildflowers* (Chambers, 1953)

BOLTON, B. *The Secret Power of Plants* (Sphere, 1975)

BOWNESS, C. *Romany Magic* (Aquarian Press, 1974)

BROUK, B. *Plants Consumed by Man* (Academic Press, 1975)

BURNE, E. *The Handbook of Folklore* (Sidgwick & Jackson, 1914)

BURNS, H. *Drugs, Medicine and Man* (Allen & Unwin, 1962)

CHAPMAN, V. J. and CHAPMAN, D. J. *Seaweeds and their Uses* (Chapman & Hall, 1980)

CLAPHAM, A. R., TUTIN, T. G. and WARBURG, E. F. *Flora of the British Isles* (CUP, 1962)

CONDRY, W. *Woodlands* (Collins, 1974)

CONNELL, C. *Aphrodisiacs in Your Garden* (Arthur Barker, 1965)

CULPEPPER, N. *Complete Herbal*, 1833 edn (Joseph Smith, 1833)

DICKINSON, C. I. *British Seaweeds, Kew series* (Eyre & Spottiswoode, 1963)

EMBODEN, W. A. jun. *Narcotic Plants* (Macmillan, 1972)

EWART, N. *The Lore of Flowers* (Blandford, 1982)

FEILD, R. *Irons in the Fire* (Crowood Press, 1984)

FITTER, R. S. R. *Finding Wildflowers* (Collins, 1971)

FOLKARD, R. *Plant Lore, Legends and Lyrics* (Sampson Low, 1892)

FREETHY, R. *The Making of the British Countryside* (David & Charles, 1981)

FREETHY, R. *The Naturalists' Guide to the British Coastline* (David & Charles, 1983)

GERARD, JOHN *The Herball or General Historie of Plantes* (London, 1636)

GRAHAM, JUDY *Evening Primrose Oil* (Thorsons, 1984)

GRIGSON, G. *The Englishman's Flora* (Phoenix House, 1958)

GRIGSON, G. *The Shell Country Book* (Phoenix House, 1962)

HARTLEY, D. *Food in England* (Macdonald, 1954)

HARTLEY, D. *Made in England* (Eyre Methuen, 1939)

HEATH, F. G. *Our British Trees and How to Know Them* (Hutchinson, 1907)

HEYWOOD, V. H. (ed.) *Popular Encyclopaedia of Plants* (CUP, 1982)

HOLBROOK, A. W. *Dictionary of British Wayside Trees* (Country Life, 1936)

HOLE C. *English Folklore* (Bell, 1940)

HUTCHINSON, J. and MELVILLE, R. *The Story of Plants and Their Uses to Man* (Waverly, 1948)

INGLIS, B. *Fringe Medicine* (Faber, 1964)

JACOB, D. *A Witches' Guide to Gardening* (Paul Elek, 1964)

JOHNS, REVD C. A. *Bushes, Trees and Shrubs* (Routledge, undated)

JOHNS, REVD C. A. *Flowers of the Field* (Routledge, 1880)

JONES, J. L. *Crafts from the Countryside* (David & Charles, 1975)

JORDAN, M. *A Guide to Wild Plants* (Millington, 1976)

KINGSBURY, J. M. *Deadly Harvest* (Allen & Unwin, 1967)

KNOCK, A. G. *Willow Basket Work* (Dryad, 1970)

KREIG, M. B. *Green Medicine* (Harrap, 1965)

LANCASTER, R. *In Search of the Wild Asparagus* (Michael Joseph, 1983)

LAUNERT, E. *Hamlyn Guide to Edible and Medicinal Plants of Britain and Northern Europe* (Hamlyn, 1981)

LOUDON, J. C. *Arboretum et fruticetum britannicum* (London, 1838)

LOVELOCK, YANN *The Vegetable Book* (Allen & Unwin, 1972)

MABEY, R. *The Common Ground* (Hutchinson, 1980)

MABEY, R. *Plants with a Purpose* (Collins, 1977)

MAJOR, A. *The Book of Seaweeds* (Gordon & Cremonesi, 1977)

MAPLE, E. *The Secret Lore of Plants and Flowers* (Hale, 1980)

MESSÉGUÉ, M. *Health Secrets of Plants and Herbs* (Collins, 1979)

MEZ-MANGOLD, P. *A History of Drugs* (Hoffman-La Roche, 1971)

MORTON, A. G. *History of Botanical Science* (Academic Press, 1981)

MOSSMAN, K. *The Shell Guide to Rural Britain* (David & Charles, 1979)

NEVILLE-HAVINS, P. J. *The Forests of England* (Hale, 1976)

NORTH, P. *Poisonous Plants and Fungi* (Blandford, 1967)

POLSON, C. J. and TATTERSALL, R. N. *Clinical Toxology* (Pitman, 1969)

PORTEOUS, A. *Forest Folklore* (Allen & Unwin, 1968)

ROSE, F. *The Wildflower Key* (Warne, 1981)

ROWE, W. H. *Our Forests* (Faber, 1946)

RUTHERFORD, MEG *A Pattern of Herbs* (Allen & Unwin, 1975)

SALISBURY, SIR EDWARD *Weeds and Aliens* (Collins, 1961)

SAVAGE, D. S. *The Cottager's Companion* (Peter Davis, 1975)

SEAGAR, H. W. *Natural History in Shakespeare's Time* (Eliot Stock, 1896)

SEYMOUR, JOHN *The Forgotten Arts* (Dorling Kindersley, 1984)

TAMPION, JOHN *Dangerous Plants* (David & Charles, 1977)

TAYLOR, W. L. *Forests and Forestry in Great Britain* (Crosby Lockwood, 1946)

TAYLOR, N, *Plant Drugs that Changed the World* (Allen & Unwin, 1965)

TURNER, W. *The Herbal* (London, 1568)

TUSSER, THOMAS *Five Hundred Points of Good Husbandry* (1580; OUP, 1984)

TUTIN, T. G. and HEYWOOD, V. H. (eds) *Flora Europaea* (CUP, vols 1–5, 1964–80)

WATSON, L. *Supernature* (Hodder & Stoughton, 1973)

WHITLOCK, R. *Historic Forests of England* (Moonraker Press, 1970)

WILKINSON, G. *Trees in the Wild* (Hope, 1973)

WILKS, J. H. *Trees of the British Isles in History and Legend* (Muller, 1972)

Index